# Plant Tissue Culture

# 植物组织培养

Editor-in-Chief Wang Yuanhua Wang Shan

主 编 王媛花 王 姗

中国财富出版社有限公司

图书在版编目（CIP）数据

植物组织培养 = Plant Tissue Culture：英文 / 王媛花，王姗主编 . —北京：中国财富出版社有限公司，2023.12

ISBN 978-7-5047-8037-9

Ⅰ . ①植… Ⅱ . ①王… ②王… Ⅲ . ①植物组织—组织培养—教材—英文 Ⅳ . ① Q943.1

中国国家版本馆 CIP 数据核字（2023）第 238064 号

| | | | | | |
|---|---|---|---|---|---|
| **策划编辑** | 刘静雯 | **责任编辑** | 刘静雯 | **版权编辑** | 李 洋 |
| **责任印制** | 尚立业 | **责任校对** | 张营营 | **责任发行** | 敬 东 |

| | | | |
|---|---|---|---|
| **出版发行** | 中国财富出版社有限公司 | | |
| **社　　址** | 北京市丰台区南四环西路188号5区20楼 | **邮政编码** | 100070 |
| **电　　话** | 010-52227588 转 2098（发行部） | | 010-52227588 转 321（总编室） |
| | 010-52227566（24小时读者服务） | | 010-52227588 转 305（质检部） |
| **网　　址** | http: //www. cfpress. com. cn | **排　　版** | 宝蕾元 |
| **经　　销** | 新华书店 | **印　　刷** | 北京九州迅驰传媒文化有限公司 |
| **书　　号** | ISBN 978-7-5047-8037-9/ Q · 0008 | | |
| **开　　本** | 710mm × 1000mm 1 /16 | **版　　次** | 2024年9月第 1 版 |
| **印　　张** | 13.25 | **印　　次** | 2024年9月第 1 次印刷 |
| **字　　数** | 231千字 | **定　　价** | 49.80 元 |

# 编写人员名单

主　编：王媛花　王　姗

副主编：张　更　许建民　王　瑞

编　者：（以姓氏笔画为序）

马　文　王全智　王其传　卢　贝　史　俊

冯英娜　刘淑芳　汤伟华　李　静　李秉宣

杨　瑞　杨　霞　席刚俊　樊亚敏

# Contents

# Module 1 Knowledge of Plant Tissue Culture

## Main Content

This module mainly introduces the concept, types, and characteristics of plant tissue culture. Plant cells are totipotent and can undergo dedifferentiation and redifferentiation under certain conditions to regenerate into an intact plant. After going through exploration, foundation laying, and rapid development, the plant tissue culture technique is now widely used in various aspects such as rapid propagation of plants, culture of virus-free plantlets, cultivation of new varieties, and germplasm preservation. After learning this module, students will have a preliminary understanding of the plant tissue culture technique.

## Work Tasks

### Task 1 Knowledge of Plant Tissue Culture

#### I. Concept of Plant Tissue Culture

Plant tissue culture (also known as culture *in vitro* or plant cloning) is a technical process in which plant materials (organs, tissues, cells, protoplasts, etc.) are cultured *in vitro* under suitable artificial aseptic conditions to allow them to grow, differentiate, proliferate, and regenerate into an intact plant.

All protoplasts, cells, tissues, and organs used for culture *in vitro* are collectively referred to as explants. As plant somatic cells, explants have the same genetic information as sex cells. Explants are multicellular, with both identical cells and different ones. These cell clusters, or some of them, can differentiate into plant organs or other tissues, so the callus formed by the dedifferentiation of an explant is heterogeneous. The callus is a mass of cells that are not morphologically differentiated but can undergo active division when the explant is injured or cultured *in vitro*. The cells, mostly parenchymal cells, are loosely disordered or relatively dense.

Both sex cells and somatic cells can produce a complete plant under certain conditions because they are totipotent. Plant cell totipotency refers to the fact that any plant cell with an intact nucleus possesses all the genetic information necessary to develop into an intact plant under suitable conditions. Dedifferentiation refers to the cells without the ability to division, returning to the meristematic state and dividing into undifferentiated cell clusters, i.e. calluses. Dedifferentiated cells can maintain vigorous division for a long time without differentiation under appropriate conditions. Redifferentiation refers to the transformation of dedifferentiated callus cells into cell clusters and tissues with certain structures and physiological functions, which constitute an intact plant or plant organ. To express its totipotency, a differentiated cell first undergoes dedifferentiation, followed by redifferentiation.

## II. Type of Plant Tissue Culture

Plant tissue culture can be classified in many ways. For example, by explants, it can be classified into the following six types.

1. Plant Culture

Plant culture refers to the aseptic culture of intact plant material, mostly with seeds as explants for culture (e.g. inducing *Cymbidium goeringii* seeds to germinate and grow into plantlets).

2. Organ Culture

Organ culture refers to the culture of plant organs *in vitro*, with roots, stems, leaves, flowers, fruits, and other organs of plants as explants. More precisely, root tips, root segments, shoot tips, shoot segments, blades, petioles, cotyledons, floral organs, fruits, and seeds can be used as materials for organ culture *in vitro*.

3. Tissue Culture

Tissue culture refers to the culture of plant tissues *in vitro*, in which meristem, cambium, and parenchyma can be used as the materials. It is deemed as plant tissue culture in a narrow sense.

4. Embryo Culture

Embryo culture is an *in vitro* culture, with immature or mature embryos isolated from ovules as explants.

5. Cell Culture

Cell culture refers to the culture of a single cell or a small cell cluster to generate a single-cell-derived clonal line, in which sex cells, mesophyll cells, and root tip cells can be used as the materials. Pollen culture is a kind of cell culture.

6. Protoplast Culture

Protoplast culture is an *in vitro* culture, with cell-wall-removed protoplasts as explants. Interspecific hybrids or new varieties may be obtained by protoplast fusion, i.e. somatic cell hybridization.

## III. Characteristics of Plant Tissue Culture

Through rapid development, plant tissue culture is now widely used, mainly with the following characteristics.

1. Economical, Convenient, and Efficient

Since plant cells are totipotent, a new plant can be regenerated from a single cell or a cluster of histocytes through plant tissue culture. Clones of various levels obtained in the culture, such as cells, tissue blocks, organs, or plantlets, are derived from a single piece of material with stable genetic information to preserve desirable traits of the original plant. When shoot tips, roots, shoots, leaves, cotyledons, hypocotyls, flower buds, and petals are used as explants, only a few millimeters or even less than one millimeter of them are needed for culture, reducing the cost of material in practical production. With the plant tissue culture technique, the production may be conducted on a small scale, and in a precise manner that cuts the use of manpower, material resources, and land. It is also much more convenient, economical, and refined than field production, pot culture, hydroponics, and sand culture, and avoids disruption by other organisms and microorganisms. Only with conventional asexual propagation will it take years or decades to propagate plantlets to a certain number, while with the plant tissue culture technique, tens of thousands of plantlets can be produced within 1–2 years. As production efficiency is improved, a large number of products are available in a short time. Since good effects can be obtained with less cultivation material, this technique is of great value and significance in practice. For example, for the promotion of new varieties and rejuvenation of superior varieties, especially the preservation, utilization, and development of famous, superior, special, and new varieties.

2. Controllable Culture Conditions

Plant tissue culture is tricky because the culture materials are grown and differentiated on artificial media under microclimatic environmental conditions. The components in the media, temperature, lighting intensity, wavelength of light, and photoperiod in the culture environment will not be affected by seasonal variation, diurnal variation, or climatic disaster, which is conducive to the stable production of tissue culture plantlets all year round.

3. Fast growth, Short cycle, and High Reproductive Rate

Appropriate culture conditions will be provided for different plants and materials *in vitro* during the plant tissue culture to make the materials *in vitro* grow rapidly, generally about one month for each culture cycle, greatly shortening the growth cycle of plants. Despite the need for certain facilities and equipment and the large energy consumption, high-quality plantlets up to specifications can be produced through plant tissue culture without delay as the plant materials *in vitro* can reproduce geometrically.

4. Convenient Management for Industrialized Production and Automation Control

Plant tissue culture leads to highly intensive industrialized production with facilities under manually regulated culture conditions such as temperature, lighting, humidity, nutrition, and hormones, creating conditions for automated management, with the typical characteristics of modern agriculture. Compared with pot culture and field culture, plant tissue culture saves massive drudgery such as intertillage and weeding, watering and fertilization, and pest control, cuts the use of land and is effort- and labor-saving and thus conducive to industrialized production.

## IV. Evolution of Plant Tissue Culture Technique

The use of the plant tissue culture technique goes back to the late $19^{th}$ and early $20^{th}$ centuries. This technique has basically gone through three stages of evolution: exploration, foundation laying, and rapid development.

1. Exploration Stage (early $20^{th}$ century to mid-1930s)

The theoretical basis of plant tissue culture stems from the cell theory proposed by German botanist Schleiden and zoologist Schwann from 1838–1839. This theory holds that all living things are composed of cells, which are the functional units of organisms and can only be obtained through cell division.

In 1902, German plant physiologist Haberlandt, based on the cell theory, proposed that the organs and tissues of higher plants could be continuously divided into individual cells, and assumed that *in vitro* cells had the potential to develop into an intact plant, that is, plant cells were totipotent. To confirm the idea, he cultured *in vitro* cells in Knop solution with sucrose. The palisade tissues, epidermal cells, and epidermal hairs from the leaves of plants such as *Lamium barbatum*, *Commelina communis, Eichhornia crassipes*, and *Ornithogalum caudatum* were selected as the culture materials. Unfortunately, none of the palisade cells were able to divide, although he saw the cells growing, cell walls thickening, and starch forming in the palisade cells. There are two main reasons for the failure of Haberlandt's experiment: First, the materials selected were highly differentiated cells; Second, the media used did not contain the growth hormones necessary to induce the division of mature cells. Haberlandt is a pioneer in plant tissue culture. His contribution lies not only in conducting the first experiment of *in vitro* cell culture, but also in putting forward his opinion on the role of blastochyle in tissue culture and foreseeing the adoption of nursing culture in his report—*An Experiment of Plant Cell Culture in Vitro* published in 1902.

In 1904, Hannig cultured embryos of radish and horseradish in the inorganic salt solution and sucrose solution. These embryos were fully developed and matured *in vitro* and germinated early to form plantlets. In 1922, German scholar Kotte and American scholar Robbins reported their success in root tip culture *in vitro*. This was the earliest experiment on root culture. In 1922, American scholar Knudson obtained a large number of orchid plantlets by embryo culture, overcoming the difficulty in orchid seed germination. In 1925, Laibach succeeded in culturing young embryos through the interspecific hybrid of flax, demonstrating the possibility of using embryo culture in the distant hybridization of plants. It is clear that the selection of culture materials is very important when the factors affecting the growth and differentiation of *in vitro* culture materials have not yet been clarified.

In the exploration stage of plant tissue culture research, as factors affecting the proliferation and morphogenesis of plant tissues and cells remain unclear, little progress has been made in more than 30 years of exploratory experiments in the research on plant tissue culture.

2. Foundation Laying Stage (mid-1930s to late 1950s)

By the mid-1930s, two important discoveries had been made in the field of plant tissue culture: one was the recognition of the importance of B vitamins for plant growth, and the other was the discovery that auxin is a natural growth regulator.

In 1934, White established the first clonal propagation system using tomato roots that showed vigorous growth and could be subcultured for propagation, achieving success in the *in vitro* root culture experiment for the first time. At first, the medium he used in the experiment contained inorganic salts, yeast leachate, and sucrose. In 1937, he succeeded in replacing the yeast leachate with three B vitamins (pyridoxine, thiamine, and nicotinic acid), and the new medium was later named White medium. At the same time, Gautheret (1934) found in his study of cambium tissue culture of *Salix permollis* C. Wang et C. Y. Yu that only after B vitamins and auxin were added into Knop solution containing glucose and cysteine hydrochloride can the growth of cambium tissue of *Salix perm*ollis C. Wang et C. Y. Yu be significantly enhanced. In 1939, Gautheret succeeded in the continuous culture of carrot root cambium for the first time. In the same year, White obtained similar continuously growing tissue cultures from tumor tissues of tobacco interspecific hybrids and Nobecourt also obtained them from carrot roots. Therefore, Gautheret, White, and Nobecourt together are known as the founders of plant tissue culture. The cultivation methods and media now are essentially attributed to these three scientists. Later, White published a book entitled *A Handbook of Plant Tissue Culture* in 1943, making plant tissue culture an emerging subject.

In 1953, Muir obtained the single-cell or dense-cell suspension that can be subcultured for propagation by adding the callus of marigold and tobacco into the liquid medium and breaking the callus with a reciprocating shaker. This verified Haberlandt's assumption of obtaining individual cells via tissue culture.

In 1958, Steward et al. successfully induced embryoids from the single-cell suspension for the culture of carrot root callus and grew them into complete plants (Fig. 1-1), which confirmed Haberlandt's assumption of cell totipotency for the first time through experiments and was considered as a milestone in the history of plant tissue culture research.

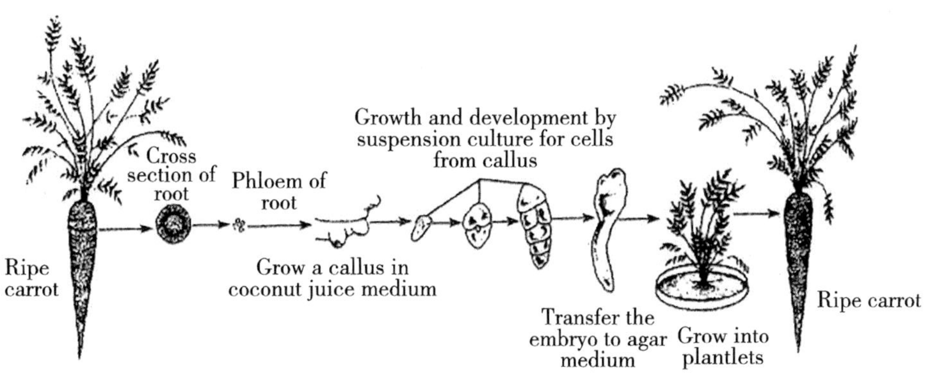


**Fig. 1-1 Schematic Diagram of Carrot Single-cell Development into Intact Plants**

(Han Yiren, 2002, *Molecular Cell Biology*)

3. Rapid Development Stage (from 1960s to now)

When the mechanism of plant cell division and organogenesis was revealed, scientists conducted a comprehensive study in this field and formed a set of proven theoretical systems and technical methods of plant tissue culture. The combination of plant tissue culture technique with conventional breeding, superior variety breeding, and genetic engineering technology has played an important role in plant improvement and has achieved considerable economic benefits. The development of the tissue culture technique since the 1960s can be summarized into four aspects.

(1) More progress in basic research: In 1965, Vasil and Hildebrandt obtained complete regenerated plants from single, isolated cells of tobacco in a medium with known chemical composition, which further confirmed the totipotency of plant cells.

(2) Breakthroughs in protoplast culture: In 1960, British scholars Cocking et al. succeeded in isolating plant protoplasts with fungal cellulase, pioneering the research of plant protoplast culture. In 1971, Japanese scholars Takebe et al. obtained regenerated plants from tobacco mesophyll protoplasts for the first time, which not only proved theoretically that in addition to somatic cells and germ cells, wall-less protoplasts also have totipotency but also provided ideal receptor materials for the introduction of exogenous genes in practice. In 1985, Fujimura obtained the first regenerated plant from the protoplast of rice, a cereal crop. In 1972, Carlon et al. obtained the first somatic interspecific hybrid plant by fusing the protoplasts of two tobacco species using sodium nitrate.

In 1974, Kao successfully obtained 3% heterocaryons by fusing the protoplasts of soybean and tobacco using polyethylene glycol (PEG). In the same year, Bonne and Eriksson extracted organelles (chloroplasts) and succeeded in the fusion of protoplasts of chloroplast-containing algae and chloroplast-free carrot roots using PEG. The results showed that protoplasts were good materials for introducing exogenous genetic materials, providing a good receptor system for genetic engineering. So far, PEG has been recognized as an ideal cell fusion agent and has been widely used in the fusion of interspecific hybrid cells with distant or no genetic relationships.

(3) Remarkable achievements in anther culture: Tulecke obtained calluses from mature pollen of ginkgo in 1953, and later from the pollen of gymnosperms such as yew, Chinese torreya, and pine. In 1964, Guha and Maheshwari successfully obtained haplobionts by pollen induction in anther culture of *Datura innoxia* Miller, thus promoting the research on anther and pollen culture. In 1967, Nitsch obtained

complete tobacco plants through the anther pre-culture. Since then, success has been achieved in the culture of tobacco, rice, wheat, corn, tomato, pepper, strawberry, apple, and other plants, totalling more than 160 species. Remarkable achievements have been made in the anther culture of tobacco, rice, and wheat in China.

(4) Wide use of micropropagation technique: In 1960, Morel proposed a method for *in vitro* asexual propagation of orchids, which features a very high multiplication coefficient. Because of its great practical value, this method was soon adopted by orchid producers, and thus the "orchid industry" quickly sprung up. In addition to orchids, the production of many other ornamental plants and cash crops (such as sugarcane and strawberry) through micropropagation has also reached an industrialized scale. For some crops, considerable economic benefits have been generated by combining micropropagation with shoot tip culture for virus elimination.

As the plant tissue and cell culture techniques are continuously improved, the research on plant genetic transformation combined with recombinant DNA technology has also become an important part of plant bioengineering. Plant genetic transformation with plant tissue culture, cell culture, and protoplast culture techniques has been successful in many plants, thus blazing a new trail and creating broad prospects for the genetic breeding of plants.

## V. Application of Plant Tissue Culture

### 1. Rapid Propagation of Plants

Rapid propagation by plant tissue culture is the most potential technique in production. This technique features a high multiplication coefficient, fast propagation speed, short propagation cycle, annual production, and uniformly sized plantlets. With this technique, a single plant can reproduce tens of thousands, hundreds of thousands or even millions of plants a year. For example, in one year, one grape plant can reproduce more than 30,000 plants; one axillary bud on an orchid stem can reproduce four million protocorms; one apical bud of a strawberry plant can reproduce $10^8$ new buds. The tissue culture technique allows the propagation of plantlets at a very fast speed, up to millions of times every year. It is particularly important for the short-time rapid propagation of some "famous, superior, special, new, and peculiar" plant species and varieties with low multiplication coefficients that cannot be propagated from seeds. Some virus-free plantlets, newly bred, newly introduced, and scarce varieties, superior single plants, endangered plants, and genetically modified plants can be rapidly propagated through tissue culture at a speed tens of thousands to millions of times faster than conventional methods, and a large number of high-quality plantlets can be obtained with this technique. Presently, industrialized plantlet raising has gradually become an important way of producing plantlets on a large scale in China and other countries (such as the industrialized production of orchids).

2. Virus-free Plantlet Culture

Due to perennial soil cultivation, some plants are infected with viruses, especially those that are asexually propagated, such as sweet potato, strawberry, potato, and garlic. With conventional propagation methods, infected mother plants will transmit viruses to the plantlets, affecting their growth and yield, and causing great economic losses in production. For example, the citrus tristeza virus once destroyed some orange orchards; the grapevine fanleaf virus reduced yield by 10%–50%. In the past, viruses were often eliminated by removing infected plants. Later, disease-resistant plants were bred and comprehensive control measures were taken. Although some achievements were made, the problem could not be eradicated because the plantlets themselves carried viruses. Production practices have proved that the use of the plant tissue culture technique can effectively eliminate the viruses from plants. Plants suitable for virus elimination generally include sweet potato, strawberry, potato, garlic, ginger, carnation, orchid, tulip, dahlia, and lily. Currently, virus-free plantlets have been obtained for more than 100 species of plants. Many production bases for virus-free plantlets have been established in China, making virus-free plantlets available for cultivation throughout the country and bringing considerable economic benefits.

3. Cultivation of New Varieties

With the plant tissue culture technique, the growth cycle can be shortened, and new varieties can be cultivated. Through protoplast fusion, it is possible to overcome the incompatibility of sexual hybridization to obtain new or superior varieties from distant hybridization; through embryo culture, the hybrid embryo develops and matures, the distant hybridization is achieved, and the breeding period is shorter; through anther culture and pollen culture, haploid breeding is achieved; in the process of cell culture, due to the influence of medium, hormones, and culture conditions, there will be some mutants, and some useful mutants may be screened to develop varieties with certain resistance or nutritional varieties. At present, short-stalked mutants with disease resistance, salt resistance, high lysine content, high protein and high yield have been screened by using this method, and some have been used for production. Cytoplasm is extracted from cells, protoplasts carrying genetic material are cultured into plantlets, or exogenous genes are introduced to breed new varieties, and homozygous and optimized combinations of multiple traits are obtained in the offspring; The breeding procedure is simplified and new varieties are developed suitable for agricultural development in the 21$^{st}$ century.

4. Germplasm Preservation

Germplasm resources are the basis of agricultural production. A significant number of plant species have disappeared or are disappearing on the earth due to natural disasters, competition between organisms and the impact of human activities on nature. It is a fact that the preservation of plant germplasm resources by plant tissue culture techniques is not affected by the environmental impact that saves space, manpower, and material resources; facilitates management, development and use as well as the exchange of germplasm resources; prevents the transmission of pests and diseases by human factors; and is conducive to the preservation and rescue of useful species genes. For example, cell suspensions for plants such as carrots and tobacco, stored at a low temperature of -196–20°C for several months, can still resume growth and develop into plants.

5. Secondary Metabolite Production

The large-scale culture of tissues or cells, combined with the fermentation technique, can produce a variety of natural organic compounds required by humans, such as proteins, fats, drugs, spices, alkaloids, and other active compounds. At present, the contents of more than 40 kinds of natural products in cultured cells are higher than those in original plants. For example, the content of ginsenoside is 21.4% in callus and 27.4% in redifferentiated plant roots, both of which are higher than that in natural ginseng roots (4.1%). More than 100 international patents have been obtained. In recent years, the research on the production of microorganisms by tissue culture methods or on drugs or active ingredients that cannot be synthesized has deepened. Some drugs or active ingredients have been used in industrial production, such as antibiotic fermentation. It is expected that there will be greater development in the future.

Under the guidance of plant cell totipotency theory, after a century of development, plant tissue culture has become an important discipline. Theoretically based on botany, plant physiology, and genetics, it involves the study of the morphogenesis, nutritional physiology, and somatic cell inheritance of plants *in vitro*. Additionally, it is an important technical means. It is not only the technical basis of plant cell engineering and gene engineering, but also an important technique for rapid propagation and virus elimination of plants. It has been widely used in agriculture, forestry, medicine, and other industries, creating great economic and social benefits.

## Task 2 Visit to Tissue Culture Factory

### I. Production Operating Process of Plant Tissue Culture

The production operating process of plant tissue culture is shown in Fig. 1-2.

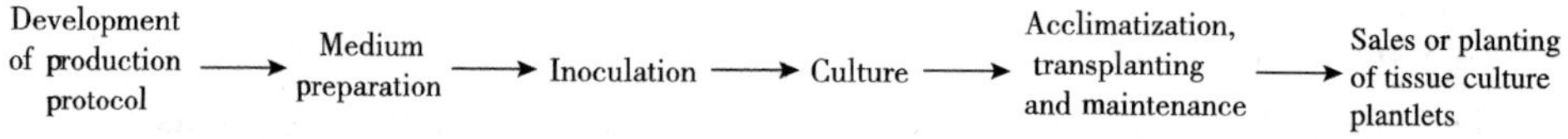


**Fig. 1-2 Production Operating Process of Plant Tissue Culture**

(Wang Zhenlong, 2014, *Tutorial on Plant Tissue Culture*)

### II. Tissue Culture Posts and Types of Work

According to the production process of plant tissue culture, a tissue culture enterprise mainly sets posts in production, research and development (R&D), management, and marketing. The production posts include medium preparation workers, inoculation workers, culture workers, as well as acclimatization, transplanting, and maintenance workers. Graduates of higher vocational colleges are mainly engaged in the production of tissue culture plantlets at first, and are allowed to engage in technical R&D and production management after three to five years of working.

## (I) Work Tasks of Tissue Culture Post

1. Medium Preparation Workers

Prepare mother liquor and working medium.

2. Inoculation Workers

Perform explant inoculation, subculture, rooting transfer, and virus elimination.

3. Culture Workers

Be responsible for the management of the culture workshop.

4. Acclimatization, Transplanting, and Maintenance Workers

Being responsible for the acclimatization, transplanting and daily maintenance of tissue culture plantlets.

## (II) Work Objectives of Tissue Culture Post

1. Medium Preparation Workers

Prepare the medium as required, accurately, normatively and skillfully; Standardize the sterilization and storage of the medium.

2. Inoculation Workers

Perform aseptic operation normatively and skillfully, control the contamination rate at 5%–10%, and eliminate viruses thoroughly.

3. Culture Workers

Perform the classified management of tissue culture plantlets as required, and the growth and differentiation should be observed carefully and recorded comprehensively. Control culture conditions depending on the culture materials.

4. Acclimatization, Transplanting, and Maintenance workers

Perform acclimatization and transplanting operations skillfully and normatively. Make sure the tissue culture plantlets have a high survival rate, are strong with good growth vigor to meet the specification requirements, and deliver the plantlets as planned.

### (III) Job Responsibilities of Tissue Culture Post

1. Medium Preparation Workers

(1) Prepare the mother liquor and medium according to operating procedures;

(2) Carefully carry out calculations, checks, and operations, and fill in and keep work records in a timely manner;

(3) Ensure that the bench surface is clean without residual liquids, the articles are arranged in a reasonable and orderly manner, and all tools and working areas are clean.

2. Inoculation Workers

(1) Keep the inoculation workshop clean;

(2) Make preparations before inoculation and strictly follow the aseptic operation procedures;

(3) Carefully keep work records;

(4) Strictly implement virus elimination protocol;

(5) Complete the work tasks to fulfill quality and quantity requirements.

3. Culture Workers

(1) Keep the culture workshop clean;

(2) Pick out contaminated plantlets and malformed plantlets in a timely manner every day;

(3) Effectively control environmental conditions according to culture needs;

(4) Observe and record the culture materials and give feedback in a timely manner;

(5) Ensure electricity safety;

(6) Identify virus-free plantlets accurately.

4. Acclimatization, Transplanting, and Maintenance Workers

(1) Keep the greenhouse clean and tidy;

(2) Operate according to the requirements for acclimatization and transplanting of tissue culture plantlets;

(3) Carry out scientific management to ensure the nutrition and environmental conditions for the growth and development of tissue culture plantlets;

(4) Carefully observe and effectively solve problems during transplanting of tissue culture plantlets;

(5) Ensure the survival rate of acclimatization and transplanting and shelf life of tissue culture plantlets.

## (IV) Requirements for Tissue Culture post

1. Knowledge and Competence

(1) Medium preparation workers

① Know the washing methods and standards of culture vessels, and can clean the glassware skillfully;

② Be skilled in preparing mother liquor and medium;

③ Be able to use the autoclave in a standardized manner;

④ Know the purpose, operating procedures, skill requirements of each link of medium preparation, and other theoretical knowledge related to medium preparation.

(2) Inoculation workers

① Identify plant organs accurately, and select and treat explants correctly;

② Select the appropriate inoculation method for different explants, and be able to carry out aseptic operation skillfully and normatively;

③ Keep aseptic awareness and know the principle of surface sterilization of explants and the methods, procedures and precautions of aseptic operation.

(3) Culture workers

① Accurately identify and effectively handle contaminated flasks and malformed plantlets;

② Know the tissue culture principles, culture conditions, common methods of tissue culture, and rapid propagation and influencing factors;

③ Be able to observe tissue culture plantlets and analyze and solve incident problems;

④ Be able to use relevant equipment to effectively control the culture environment;

⑤ Be able to implement scientific and effective management for different culture objects.

(4) Acclimatization, transplanting, and maintenance workers

① Be familiar with the purpose, principle, period, and conditions of acclimatization and transplanting of tissue culture plantlets;

② Be able to formulate scientific implementation protocol according to the acclimatization and transplanting objects, and carry out acclimatization and transplanting operations skillfully and normatively;

③ Be familiar with the characteristics, performance and use of relevant facilities and equipment, and be able to construct and maintain simple facilities for growing;

④ Have certain cultivation and maintenance abilities.

2. Quality

Dedicated, honest and trustworthy, hard-working, obedient to leadership; Abide by operating specifications and professional ethics; Works proactively and responsibly; Has cost consciousness, market consciousness, innovation consciousness, team spirit, scientific thought, strong learning ability, communication ability, planning ability, adaptability, problem-solving ability, and self-management ability.

## Summary

Plant tissue culture (also known as culture *in vitro* or plant cloning) is a process in which plant materials (organs, tissues, cells, protoplasts, etc.) are cultured *in vitro* under suitable artificial aseptic conditions so that they can grow, differentiate, proliferate, and regenerate into an intact plant. All protoplasts, cells, tissues, and organs used for culture *in vitro* are collectively referred to as explants. Plant cell totipotency refers to the fact that any plant cell with an intact nucleus possesses all the genetic information necessary to develop into an intact plant under suitable conditions.

According to the type of explants, plant tissue culture can be divided into plant culture, organ culture, tissue culture, embryo culture, cell culture, and protoplast culture.

Plant tissue culture features low cost, convenience, high efficiency, manual control of culture conditions, fast growth, short cycle, high reproduction rate, and convenient management, which is conducive to industrialized production and automation.

The development of the plant tissue culture technique can be broadly divided into three stages: exploration, foundation laying, and rapid development.

Plant tissue culture is widely used in plant rapid propagation, virus-free plantlet culture, cultivation of new varieties, germplasm preservation, and secondary metabolite production.

The production operating process of plant tissue culture: development of production scheme → medium preparation → inoculation → culture → acclimatization, transplanting, and maintenance → sales or field planting of tissue culture plantlets.

According to the production process of plant tissue culture, the tissue culture enterprise mainly has production posts, R&D posts, management posts, and marketing posts. The production posts include medium preparation workers, inoculation workers, culture workers, as well as acclimatization, transplanting, and maintenance workers.

Tissue culture plantlet factories have tissue culture plantlet production workshops and acclimatization and cultivation areas. The production workshop mainly includes the washing workshop, medium preparation workshop, sterilization workshop, changing room, inoculation workshop, culture workshop, and testing workshop. The acclimatization and cultivation area includes the transplanting and acclimatization workshop and plantlet nursery, and the plantlet nursery includes the original seed nursery, variety cultivation demonstration area, and propagation nursery.

# Review Test

## I. Explain the Glossary

1. Plant tissue Culture
2. Explant

3. Callus
4. Plant Cell Totipotency
5. Dedifferentiation
6. Redifferentiation

## II. Fill in the Blanks

1. According to the type of explants, plant tissue culture can be divided into ( ), ( ), ( ), ( ), ( ), and ( ).

2. Plant ( ) was first proposed by the German plant physiologist Haberlandt in 1902.

3. White published a book entitled *A Handbook of Plant Tissue Culture* in ( ), making plant tissue culture a new subject.

4. The development of plant tissue culture technique can be broadly divided into three stages: ( ), ( ) , and ( ).

5. In 1960, British scholar Cocking successfully isolated protoplasts using ( ), thereby pioneering the protoplast culture technique.

6. Production operating process of plant tissue culture: development of production protocol → medium preparation → ( ) → culture → ( ) → sales or field planting of tissue culture plantlets.

7. The inoculation worker is mainly responsible for ( ).

8. The objective of the medium preparation worker is to prepare the medium ( ).

9. Tissue culture plantlet factories have tissue culture plantlet ( ) and ( ).

## III. Short Answer Questions

1. What are the broad and narrow definitions of plant tissue culture?
2. What are the characteristics of plant tissue culture?

# Module 2 Plant Tissue Culture Laboratory Design and Equipment

## Work Tasks

### Task 1 Design and Composition of Plant Tissue Culture Laboratory

#### I. Design of Plant Tissue Culture Laboratory

##### (I) Overview

Before plant tissue culture is conducted, a comprehensive understanding of the most basic equipment conditions required in the work should be obtained first, so as to make use of existing houses, build a new laboratory, or rebuild the existing laboratory according to local conditions. The size of the laboratory depends on the purpose and scale of the work. For the purpose of industrialized production, a laboratory being too small will limit production and affect efficiency. In the design of the tissue culture laboratory, the tissue culture procedure should be followed to avoid inverse arrangement of some links and confusion in future work. Plant tissue culture is performed under strict aseptic conditions. To achieve aseptic conditions, certain equipment, instruments, and tools are required, and culture conditions such as temperature, photoperiod, and humidity should be manually controlled.

## (II) Design Principles of Plant Tissue Culture Laboratory

1. Site Selection

(1) Requirements for geographical location: open, high, and dry terrain, with convenient transportation and easy access to water and electricity. The environment should be clean with fresh air and there should be no sauce and vinegar factories, pasture, livestock housing, feed storage, and factories discharging "Three Wastes" within 50 meters around.

(2) Requirements for inoculation room: To reduce the contamination of undesired microbes, the inoculation room should not be too large, should be airtight if possible, and should be equipped with a buffer room in front of the inoculation operation room.

(3) Requirements for culture room: It should be built in the south of the house with heat preservation, ventilation, and sufficient light, with large windows to expose to natural light, thus saving energy.

2. Design Principles for Tissue Culture Laboratory

(1) Design should prevent and control contamination and ensure aseptic operation.

(2) Design should be made scientifically according to the process flow to achieve low cost, practicality and high efficiency.

(3) Structural layout should be reasonable for the purpose of convenient work, energy saving, and safety.

(4) The planning and design should be suitable for the work purpose, scale, and local conditions.

## II. Composition of Plant Tissue Culture Laboratory

The area of the plant tissue culture laboratory should be determined according to the category of work and the production scale. A small area is mainly used for scientific research and small-scale production, while a large area is designed for large-scale industrialized production, which is also commonly called a "Tissue Culture Factory". Except for the difference in scale, the construction requirements, structures, and functions are basically the same for both small laboratories and large tissue culture factories.

A large tissue culture laboratory includes a washing room, drug room, weighing room, medium preparation room, sterilization room, inoculation room, culture room, observation and recording room, storage room, as well as a certain area of plantlet transfer greenhouse outdoors. Small laboratories can be made by merging similar parts. For example, a washing room, medium preparation room, and sterilization room can be incorporated into a preparation room; a drug room, weighing room, and observation and recording room can be incorporated into a complex laboratory. However, the inoculation room, culture room, and plantlet transfer room should be set up separately.

The laboratory layout should be reasonable. Generally, a continuous production line is arranged according to the working procedure of tissue culture to avoid increased work amount or confusion in the future due to the inverse arrangement of some links. It should be designed in a way that is convenient for working, reduces contamination, and saves energy. It should also be equipped with fire-fighting facilities to ensure safety of use. Fig. 2-1 shows the room layout of a small plant tissue culture laboratory.

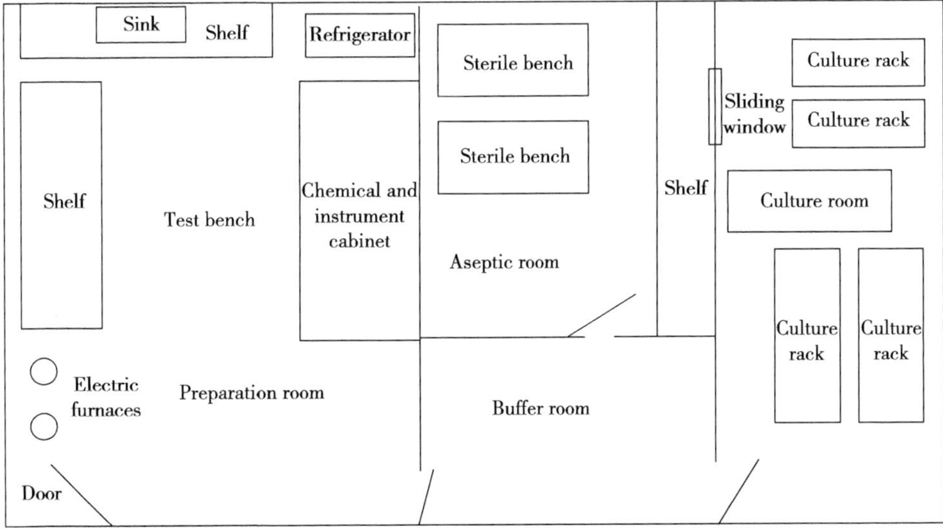


**Fig. 2-1 Room Layout of a Small Plant Tissue Culture Laboratory**

1. Preparation Room (Chemical Laboratory)

Requirement: The preferable area is about 20 $m^2$. It should be spacious and bright to accommodate multiple test benches and related equipment for several people working at the same time. It should be well-ventilated for gas exchange, and the floor should be easy to clean and should be subjected to anti-skid treatment.

2. Washing and Sterilization Room

Function: Complete washing, drying, preservation of various tools, and sterilization of medium.

Equipment: Sink, dripping rack, central test bench, autoclave, ultrasonic cleaner, drying sterilizer (e.g. drying oven), etc.

Requirement: The size of the washing and sterilization room is determined according to the work amount, generally 30–50 $m^2$. A dedicated washing tank should be set on one side of the laboratory to clean the glassware. The central test bench should also be equipped with two tanks to clean small glassware. One or two washing tanks should be set to wash the glassware with high requirements on cleanliness. The floor should be resistant to wet and well-drained.

3. Buffer Room

Function: It is a buffer site before entering the aseptic room to reduce dust and other pollutants brought by people from the outside. Staff can wear work clothes, slippers, and masks here before entering the aseptic room and culture room. Before entering the aseptic operation room, staff should change clothes and shoes here to reduce undesired microbes brought into the inoculation room.

Equipment: One or two ultraviolet lamps, sink, test bench, shoe and hat rack, cabinet, sterilized work clothes, slippers, and masks.

Requirement: The buffer room should be 3–5 $m^2$ and can be built outside the preparation room or outside the aseptic operation room. It should be kept clean and sterile, provided with a shoe rack, hat-and-coat hook, clean slippers for the laboratory, and sterilized work clothes. One or two ultraviolet lamps should be installed at the top of the wall to regularly sterilize clothes. An ultraviolet lamp for radiation sterilization should also be installed in the buffer room. The door of the buffer room should be staggered with the door of the inoculation room, and the two doors should not be opened at the same time so as to ensure that no undesired microbes can be brought into the aseptic room due to door opening and staff entry and exit.

4. Aseptic Operation Room (Inoculation Room)

Function: It is mainly used for disinfection and inoculation of plant materials, transfer of culture, subculture of test-tube plantlets, preparation of protoplasts, and all technical procedures to be performed under aseptic conditions. It is the most critical step in plant *in vitro* culture research or production.

Equipment: Ultraviolet lamp, air conditioner, dissecting microscope, sterilizer, alcohol burner, inoculation apparatus (inoculation tweezers, scissors, scalpel, and inoculation needle), test bench, and shelf, as well as a clean bench and a complete set of sterilization and inoculation instruments and chemicals.

Requirement: The inoculation room should be small, generally 10–20 $m^2$. It should be well-closed, dry, quiet, clean, bright, and keep sterile for a long time, so it should not be set in a place prone to moisture. The floor, ceiling and walls should be as well-closed and smooth as possible, easy to clean and disinfect, and have sliding doors to reduce the air disturbance when opening and closing doors. To facilitate disinfection, the floors and inner walls should be waterproof and corrosion-resistant. To keep the aseptic room clean, cross-ventilation is not allowed. The inoculation room requires one to two ultraviolet lamps to be mounted at proper positions for irradiation sterilization. A small air conditioner should be installed to control the indoor temperature, so that the doors and windows can be closed tightly, reducing cross-ventilation with the outside.

5. Culture Room

Function: It is used to culture the plant *in vitro* materials (either research experimental or production experimental materials), inoculated into culture flasks and other vessels under controlled conditions.

Equipment: Culture rack (temperature, illumination, and humidity controlled), shaker, rotating bed, automatic timer, ultraviolet lamp, light incubator or artificial climate chamber, biochemical incubator, side test bench for photographing culture growth, dehumidifier, microscope, hygrothermograph, air conditioner, etc.

Requirement: The culture room should be 10–20 $m^2$. The size of the culture room can be determined according to the size and number of culture racks and other auxiliary equipment to make full use of space and save energy. The basic requirement is to control illumination and temperature and maintain a relatively sterile environment. Air exchange devices such as exhaust windows and ventilation fans should be installed in the in-culture room to meet the air needs for the growth of plant culture materials. Appropriate culture racks should be provided and fluorescent lamps should be installed to save energy and space.

6. Observation and Recording Room

It is designed to observe and record the growth of cultures and experimental results. It should have a fixed terrazzo bench to place the microscope, dissecting microscope, microscope camera, and other instruments. It is better to include a set of slide preparations and cytological staining equipment for slide preparation or staining. The room should be quiet, clean, and bright, and the instruments should not vibrate or get wet.

7. Storage Room

It is designed to store temporarily unused vessels and tools. It should be well-ventilated, preferably located on a low floor on the shady side of the building to facilitate the handling and storage of articles.

8. Plantlet Transfer Room

It is designed for the acclimatization and transplanting of test-tube plantlets. It should be temperature controlled, moisture retained, protected from direct sunlight, generally kept at 15–25°C, and the relative air humidity above 70% and protected from direct sunlight. Common greenhouse or plastic greenhouse can be used after appropriate renovation, and be equipped with a mist sprayer, sunshade net, insect-proof screen, transplant bed, nutrition bowl, transplant medium, etc.

## Task 2 Equipment and Instruments Commonly Used in Plant Tissue Culture

1. Balance

Two to three balances with different accuracy are required in the tissue culture laboratory. A balance with a sensitivity of 0.001 g (analytical balance) and a balance of 0.0001 g (electronic balance) are used to weigh trace elements and some experimental supplies requiring higher accuracy. Balances with sensitivities of 0.01 g and 0.1 g are used to prepare major element mother liquor and weigh some chemicals used in large amounts.

2. Refrigerator

Cryopreservation is required for various vitamins, hormones, and medium mother liquor. For some tests, plant materials should be treated at a low temperature, generally in common refrigerators.

3. Acidimeter

It is used to measure the pH value of medium and other solutions, generally within the range of 1–14, with an accuracy of 0.01.

4. Centrifuge (Fig. 2-2)

It is used for the separation of cells and protoplasts, as well as the separation and extraction of organelles, nucleic acid, and protein from cultured cells. Different types of centrifuges are used according to the separated substances. For example, low-speed centrifuge for the separation of cells and protoplasts, high-speed refrigerated centrifuge for the separation of nucleic acid and protein, and large-scale centrifugal separation system for large-scale production of secondary products.

**Fig. 2-2 Centrifuge**

5. Heater

It is used for medium preparation. Heaters with magnetic stirring functions are generally used in research laboratories, while high-power heating and electric stirring systems are used in large-scale laboratories.

6. Medium filling device

The simple and convenient equipment: It is a "bucket" bought from the medical device store, with a segment of hose tube and a spring water stopper fixed on the lower tube. A small amount of medium can also be filled directly with a funnel, and an automatic quantitative filling machine can be considered in the case of large-scale or higher efficiency requirements.

7. Sterilization Equipment

The essential requirement for tissue culture laboratories is to maintain a relatively sterile environment.

(1) Autoclave (Fig. 2-3)

It is used for the sterilization of medium, glassware, and other articles available for high-temperature sterilization. It is divided into different specifications according to the size, including portable type, vertical type, horizontal type, etc.

**Fig. 2-3 Autoclave**

(2) Dry heat sterilizer

It is used for the sterilization of metal tools such as forceps, scissors, scalpels, and glassware. Generally, the common or far-infrared sterilizer at about 200°C is selected.

(3) Filter sterilizer

It is used to sterilize some enzyme preparations, hormones, vitamins, and other non-autoclaving reagents, mainly including the vacuum filtration type and the syringe type.

(4) Ultraviolet lamp

It is a convenient and cost-effective device for controlling the sterile environment and is essential for the buffer room, inoculation room, and culture room.

8. Inoculation Equipment

(1) Clean bench (Fig. 2-4)

It is widely used for disinfection, cutting, separation, and transfer of culture materials and is the most commonly used and popular aseptic operating equipment for plant tissue culture. It is easy and comfortable to use, featuring high work efficiency, good sterilization effect, and short preparation time.

**Fig. 2-4 Clean Bench**

The clean bench is divided into single-person, double-person and three-person types, as well as open and closed types, and is composed of operation area, fan room, air filter, lighting facilities, etc. During operation, the pre-filtered air is sent to the plenum chamber with the help of the fan. Then, the dust, bacterial, and fungal spores greater than 0.3 μm in the air are removed through the HEPA filter, and sent out in the form of vertical or horizontal laminar flow, thus achieving high cleanliness in the operation area. In this way, a relatively sterile environment is formed on the bench surface, which can effectively reduce the contamination of undesired microbes and greatly improve the success rate of inoculation.

The clean bench is generally placed in the aseptic room and the room is kept clean to prolong the service life of the filter. The bench is generally wide, so attention should be paid when purchasing and designing a room to remain enough width to let the bench in successfully. When the clean bench is used too long, the filter device may be blocked, the air speed may be reduced, and the aseptic operation cannot be ensured; the filter needs to be cleaned or replaced.

(2) Inoculation hood

To save cost, the inoculation hood can be used instead of the clean bench. The inoculation hood is a well-closed wooden or glass box, which ensures the sterility of the internal space through sealing, agent fumigation, and ultraviolet lamp irradiation. However, it has some disadvantages, such as limited operation activities, long preparation time and low work efficiency.

(3) Inoculation tool sterilizer

It is used to sterilize inoculation tools. It is made of stainless steel, including horizontal and vertical types. It is efficient with built-in heating elements and numerical display temperature control.

(4) Dissecting microscope

It is used to strip the shoot tip. For easy observation and operation, a binocular dissecting stereoscope can be used to separate small tissues such as the shoot tip, usually magnified by 5–80 times.

9. Culture Equipment

(1) Culture rack

As a common facility for plant propagation and culture in all plant tissue culture laboratories, it is cost-effective, flexible and easy to use while making full use of the culture space. It is generally divided into four to five layers, with an interval of 40–50 cm between layers. The lighting intensity can be determined based on the characteristics of the cultured plants. Generally, two to four fluorescent lamps are equipped for each rack.

(2) Incubator

For experiments requiring accurate culture such as cell culture and protoplast culture, a light incubator, $CO_2$ incubator, and humidity control incubator can be used.

(3) Shaker

To improve ventilation during liquid culture, an oscillating incubator or shaker can be used.

10. Other Equipment

Time-programmed controller, temperature control system or air-conditioned stereomicroscope, inverted biological microscope as well as the corresponding photography, video recording, and image processing equipment; Electrophoresis apparatus, extraction and chromatography equipment, ultraviolet spectrophotometer, as well as HPLC, GC, and ELISA system.

11. Vessels and Tools

Culture vessels refer to sterile devices that contain medium and provide space for culture growth. Culture tools refer to various metal or plastic products used for culture inoculation and sealing, which are necessary for the laboratory.

(1) Culture containers

Various specifications of culture dishes, triangular flasks, test tubes, and culture flasks; glass containers (generally made of colorless borosilicate glass without color refraction); plastic containers (light, not easy to break, easy to make, and widely used, generally made of polypropylene and polycarbonate).

(2) Metal tools

① Tweezers.

Small pointed tweezers are used for plant tissue dissection, and gun-type tweezers, 16–22 cm long, are used for transfer and propagation of ramets.

② Scissors.

Stainless steel medical curved-head scissors, sampling and cutting for transfer and propagation, 14–22 cm.

③ Scalpel.

Medical stainless steel scalpel with a short handle and replaceable blade for tissue cutting.

④ Other tools.

Inoculating shovels and needles are useful in anther and pollen culture.

(3) Plastic tools

Various sealing films, covers, plastic discs, and other experimental tools.

## Summary

To sum up, we mainly introduce the design and composition of plant tissue culture laboratory as well as the commonly used instruments, vessels and tools in the laboratory. After learning this module, students will master the basic design requirements, composition, and functions of each part of the plant tissue culture laboratory, and understand and master the use and maintenance of the main equipment, vessels, and tools used in the laboratory, so as to lay a foundation for further study of plant tissue culture techniques.

## Review Test

1. Design a plan for constructing a plant tissue culture laboratory according to what you have learned and inquired about.

2. What are the main parts of the plant tissue culture laboratory? What functions do they have?

3. What is the necessary equipment for a plant tissue culture laboratory?

# Module 3 Basic Operation of Plant Tissue Culture

## Main Content

This module mainly introduces the general procedures of plant tissue culture, the preparation methods of medium mother liquor and medium, the selection and disinfection of explants, the primary culture of explants, the subculture of cultures, and the control of culture conditions. It also analyzes the causes of contamination, browning, and vitrification during the culture process, proposes control measures and solutions, and describes the screening methods of the optimal culture scheme. After learning this module, students will master the basic operating techniques and principles of each link of tissue culture to solve the technical problems in each link.

## Work Tasks

### Task 1 Preparation of MS Medium Mother Liquor

As an important substrate and material basis of plant tissue culture, the medium is one of the key factors in determining the success of tissue culture. The nutrients required for plant tissue culture are mainly obtained from the medium, most of which include water, inorganic salts, organic matter, plant growth regulators, coagulants, and other additives.

The medium is divided into minimal medium and complete medium according to the nutrient level. Minimal medium refers to the medium only containing such nutrients as inorganic salts, organic matter, vitamins, inositol, and amino acids, i.e. MS, $B_5$, $N_6$, and other media. Complete medium refers to the medium that can be directly used for tissue culture by adding appropriate plant growth regulators, organic additions and coagulants to the minimal medium.

The medium preparation is the most basic task in plant tissue culture. However, it is difficult to prepare as it contains more than a dozen chemicals and requires a small amount of trace elements and phytohormones, which should be weighed in each preparation, making the preparation time-consuming, laborious, and inaccurate. Therefore, for easy use and accurate amount, major elements, trace elements, iron salts, organic matter, and hormones are often separately prepared into mother liquor (i.e. stock solution with higher concentration) several folds thicker than the required concentration in the medium formulation. In the medium preparation, it is only necessary to pipette a precalculated amount of mother liquor. Let's take MS medium as an example to introduce the preparation of mother liquor.

## I. Preparation of MS Medium Mother Liquor

According to the medium formulation, prepare the chemicals and plant growth regulators used for the preparation of MS medium mother liquor, and calculate the quantity of each chemical weighed and the quantity pipetted for 1 L medium.

**Knowledge Points**

In addition to meeting the conditions of basic experimental equipment, students must also master the basic operation of plant tissue culture, which is generally divided into the following steps.

To prepare the medium rapidly and ensure the accuracy of each component and the rapid use for the preparation, the nutrient elements in the medium are generally prepared into a concentrated solution, i.e. mother liquor. According to the type and chemical properties of nutrients, the mother liquor generally includes major element mother liquor, trace element mother liquor, iron salt mother liquor, organic matter mother liquor, and growth regulator mother liquor. The mother liquor preparation process is shown is Fig. 3-1.

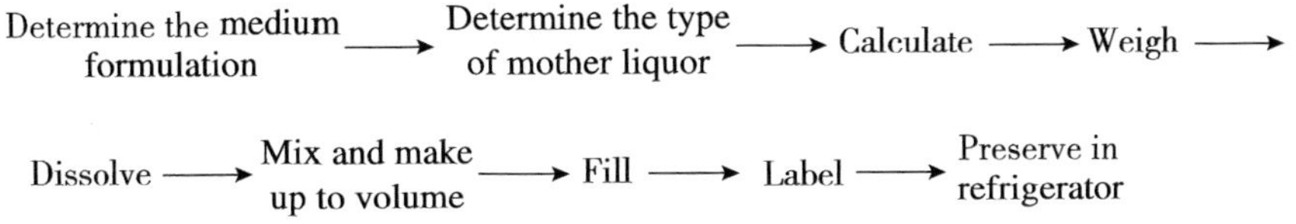


**Fig. 3-1 Mother Liquor Preparation Process**

The concentration factor of each mother liquor is slightly different, generally 10–20 folds for major elements, 100–200 folds for trace elements, 100 folds for iron salts, and 100–200 folds for organic matter. Generally, the concentration of plant growth regulator mother liquor is 0.1–1 mg/mL, and 50 mL or 100 mL is prepared at one time.

## II. Calculation

Amount of chemical for preparation of MS mother liquor (mg) = Formulation amount (mg/L) × Concentration factor × Preparation volume (L)

Amount of chemical for preparation of plant growth regulator mother liquor (mg) = Concentration of mother liquor (mg/mL) × Volume of mother liquor (mL)

## III. Preparation Process

1. MS Medium Mother Liquor

Generally, a balance with a sensitivity of 0.01 g is used for weighing major elements and iron salts, and a balance with a sensitivity of 0.0001 g is used for weighing trace elements and organic matter. Weigh the chemicals according to Table 3-1 and add them into beakers containing 600–700 mL of distilled water, respectively, stir gently with a glass rod to dissolve them, then pour them into a 1,000 mL volumetric flask, rinse the beakers 3–4 times with distilled water, add water to volume, and shake well; Pour the prepared mother liquor into reagent bottles, affix a label, indicate the name of medium, name of mother liquor, concentration factor, preparation date, preparation personnel, and then store them in a 4°C refrigerator.

**Table 3-1 Preparation of MS Medium Mother Liquor**

| Name of Mother Liquor | Component | Prescribed Amount (mg/L) | Concentration Factor | Weighed Amount (mg) | Volume of Mother Liquor (mL) | Volume Pipetted for 1 L Medium (mL) |
|---|---|---|---|---|---|---|
| Major elements | $KNO_3$<br>$NH_4NO_3$<br>$MgSO_4 \cdot 7H_2O$<br>$KH_2PO_4$<br>$CaCl_2 \cdot 2H_2O$ | 1,900<br>1,650<br>370<br>170<br>440 | 10 | 19,000<br>16,500<br>3,700<br>1,700<br>4,400 | 1,000 | 100 |
| Trace elements | $MnSO_4 \cdot 4H_2O$<br>$ZnSO_4 \cdot 7H_2O$<br>$H_3BO_3$<br>KI<br>$Na_2MoO_4 \cdot 2H_2O$<br>$CuSO_4 \cdot 5H_2O$<br>$CoCl_2 \cdot 6H_2O$ | 22.3<br>8.6<br>6.2<br>0.83<br>0.25<br>0.025<br>0.025 | 100 | 2,230<br>860<br>620<br>83<br>25<br>2.5<br>2.5 | 1,000 | 10 |
| Iron salts | $Na_2EDTA$<br>$FeSO_4 \cdot 4H_2O$ | 37.3<br>27.8 | 100 | 3,730<br>2,780 | 1,000 | 10 |
| Organic matter | Glycine<br>Thiamine hydrochloride ($VB_6$)<br>Pyridoxine hydrochloride ($VB_1$)<br>Nicotinic acid<br>Inositol | 2.0<br>0.1<br>0.5<br>0.5<br>100 | 100 | 200<br>10<br>50<br>50<br>10,000 | 1,000 | 10 |

When preparing iron salt mother liquor, weigh ferrous sulfate tetrahydrate ($FeSO_4 \cdot 4H_2O$) and ethylene diamine tetraacetic acid ($Na_2EDTA$) using an electronic balance with a sensitivity of 0.0001 g, respectively heat them with distilled water, boil to completely dissolve, mix after dissolution, then continue boiling for several minutes (heat for about five minutes until the solution turns golden yellow), cool and then make up to volume in a 1,000 mL volumetric flask, and shake well, pour into brown reagent bottle, place at room temperature for a period of time until full reaction, affix a label, give clear indication of the name of the medium, name of mother liquor, concentration factor, preparation date, and preparation personnel. Then store in a 4°C refrigerator.

2. Plant Growth Regulator Mother Liquor

Each plant growth regulator mother liquor must be separately prepared. According to Table 3-2, weigh the corresponding chemicals using a balance with a sensitivity of 0.001 g, and add a small amount of the corresponding dissolving agent to dissolve. After fully dissolving, rinse the beaker with distilled water three to four times, add water to volume, shake well, then pour into a reagent bottle, affix a label, indicate the name of mother liquor, concentration, preparation date and preparation personnel, and then store in a 4°C refrigerator.

**Table 3-2 Preparation of Growth Regulator Mother Liquor**

| Name of Mother Liquor | Chemical | Dissolution Medium | Concentration of Mother Liquor (mg/mL) | Volume of Mother Liquor (mL) | Weighed Amount (mg) |
|---|---|---|---|---|---|
| NAA mother liquor | NAA | 95% alcohol, 1 mol/L NaOH | 0.1 | 100 | 10 |
| IBA mother liquor | IBA | 95% alcohol, 1 mol/L NaOH | 0.1 | 100 | 10 |
| 6-BA mother liquor | 6-BA | 1 mol/L HCl, 1 mol/L NaOH | 0.1 | 100 | 10 |
| 2,4-D mother liquor | 2,4-D | 95% alcohol, 1 mol/L NaOH | 0.1 | 100 | 10 |

## Knowledge Points

1. Knowledge about MS Mother Liquor Preparation

Chemicals used for mother liquor preparation should be analytically pure or chemically pure to prevent precipitation. Tracing paper should be used for weighing to prevent chemicals from being stained on the paper, thus affecting the accuracy; Special keys should be used for special chemicals to avoid mix-up; when it is close to the weighed amount, gently beat the arm to prevent excessive amount. The weighed chemicals should be marked to prevent missed weighing or repeated weighing.

Distilled water should be used to dissolve the chemicals. The distilled water 60%–70% of the preparation volume, should be added into the beaker, and the weighed chemicals should be added in sequence. A chemical can be added after the previous is completely dissolved until all the chemicals in the mother liquor are added to prevent precipitation. For insoluble chemicals, they can be heated at a temperature of 60–70℃. Note that $Ca^{2+}$, $SO_4^{2-}$, and $PO_4^{3-}$ are prone to precipitation when the major element mother liquor is prepared, so $CaCl_2 \cdot 2H_2O$ should be added after $MgSO_4 \cdot 7H_2O$ and $KH_2PO_4$ are fully dissolved.

In the preparation of iron salt mother liquor, as iron ions are unstable in the aqueous solution and are easy to react with hydroxide ions to precipitate crystals, it must be prepared separately. During the preparation process, it is necessary to heat and fully dissolve the mixture, respectively, and then mix it for stable chelation. Since it is easy to decompose under light exposure, it should be stored in a brown reagent bottle at a low temperature.

2. Knowledge about Plant Growth Regulator

The plant growth regulator is the key substance in the medium. Although only a small amount is required, it plays an important role. The type, concentration, and proportion of plant growth regulator can be determined according to the purpose of tissue culture, the type of explant, organs, and the growth performance. It is used to regulate the growth and development process, differentiation direction and organogenesis of plant tissues. As a key to a good medium, it plays a decisive role in plant tissue culture. Plant growth regulators include auxin, cytokinin, gibberellin, abscisic acid, paclobutrazol, and others. The most commonly used ones in plant tissue culture are auxin and cytokinin.

(1) The main function of auxin is to promote cell division and elongation, induce the production of calluses, promote the rooting of shoot tip, and induce the formation of adventitious embryos in some plants. The commonly

used auxins include indoleacetic acid (IAA), naphthylacetic acid (NAA), indolebutyric acid (IBA), 2,4-dichlorophenoxyacetic acid (2,4-D), etc. Auxin and cytokinin are used in combination to promote the differentiation of adventitious buds as well as the germination and growth of lateral buds. However, 2,4-D often inhibits the formation of buds. It features a narrow appropriate amount range, and is toxic in excessive amounts, so it is generally used in cell initiation and dedifferentiation stages, while IAA, NAA, and IBA are generally used in differentiation induction and proliferation stages. The intensity of actions is in the order of 2,4-D > NAA > IBA > IAA. The commonly used concentration is 0.1–10 mg/L. Auxin is generally dissolved in 95% alcohol or 1 mol/L NaOH solution, and it is dissolved better in the latter. 2,4-D should be stored in brown reagent bottles.

(2) The primary function of cytokinin is to promote cell division and expansion, induce the formation of embryoids and adventitious buds, delay tissue aging, and promote protein synthesis. In the plant tissue culture, the ratio of cytokinin to auxin controls the organ development pattern. A higher concentration of auxin will promote root formation, while a higher concentration of cytokinin will promote bud differentiation. Common cytokinins include 6-benzylaminopurine (6-BA), kinetin (KT), and zeatin (ZT). Cytokinins are usually soluble in 1 mol/L NaOH, 95% ethanol, or 1 mol/L HCL.

(3) Other plant growth regulators.

The main function of gibberellin (GA) is to promote the elongation and growth of plantlet stems. GA has an inhibitory effect on the formation of plant organs and embryoids. Never the less, after organ formation, it can promote the growth of organs or embryoids as well as the development of adventitious embryos into plantlets. In addition, gibberellin and auxin act synergistically on cambium differentiation. A higher ratio of auxin to gibberellin is conducive to xylem differentiation, while a lower ratio is conducive to phloem differentiation.

Abscisic acid (ABA) can promote the formation and development of somatic embryos, mature the somatic embryos while not germinating, and promote the differentiation of some adventitious buds.

Paclobutrazol ($PP_{333}$) can control dwarfing, promote branching, grow roots and flowers, delay aging, and enhance plant stress resistance. It is mainly used for the strong growth and rooting of test-tube plantlets in plant tissue culture to improve stress resistance and the survival rate of transplanting.

The mother liquor stored in a 4℃ refrigerator should be checked regularly for precipitation and mold, and it should be re-prepared in case of any precipitation and mold.

## Task 2 Preparation of MS Medium

### Knowledge Points

1. Process of Medium Preparation (Fig. 3-2)

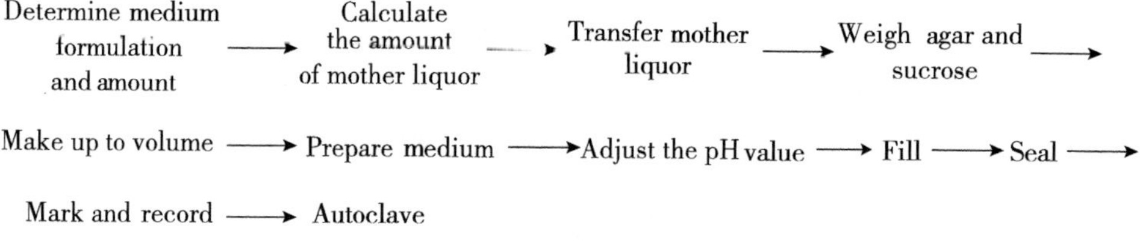


**Fig. 3-2 Medium Preparation Process**

2. Determine the Medium Formulation

The medium formulation should be determined according to the culture object and culture purpose by consulting the data. The amount of medium should be determined according to the number of inoculated explants and the number of test treatments. Take this formulation as an example: 1 L MS + 0.5 mg/L NAA + 2 mg/L 6-BA+ 3% sucrose + 0.7% agar, with a pH value of 5.8, and introduce the preparation of MS medium.

## I. Calculation

Calculate the amount of various mother liquors and fill the calculated data in the Table 3-3.

$$\text{Amount of MS mother liquor (mL)} = \frac{\text{Volume of prepared medium (mL)}}{\text{Concentration factor of mother liquor}}$$

$$\begin{array}{c}\text{Amount of plant growth}\\ \text{regulator mother liquor (mL)}\end{array} = \frac{\text{Volume of prepared medium (mL)}}{\text{Concentration factor of mother liquor}} \times \begin{array}{c}\text{Medium preparation}\\ \text{volume}\end{array}$$

$$\text{Amount of sucrose and agar weighed} = \begin{array}{c}\text{Percentage}\\ \text{concentration}\end{array} \times \begin{array}{c}\text{Medium preparation}\\ \text{volume}\end{array}$$

**Table 3-3 Preparation of MS Solid Medium**

Prepared by: Prepared on:

| Name of Medium Mother Liquor | Concentration Factor of Mother Liquor (Concentration) | Preparation Volume (mL) | Amount to be Taken or Weighed this Time |
|---|---|---|---|
| Major element mother liquor | 10 | 1,000 | 100 mL |
| Trace element mother liquor | 100 | 1,000 | 10 mL |
| Iron salt mother liquor | 100 | 1,000 | 10 mL |
| Organic matter mother liquor | 100 | 1,000 | 10 mL |
| NAA | 0.1 mg/mL | 1,000 | 5 mL |
| 6-BA | 0.2 mg/mL | 1,000 | 2.5 mL |
| Sucrose | 3% | 1,000 | 30 g |
| Agar | 0.7% | 1,000 | 7 g |

## II. Measuring

Add 30% of the medium volume of distilled water into the beaker, according to the calculated amounts of mother liquor. Add the mother liquors of the major element, trace element iron salt, organic matter, plant growth regulator, and sucrose into the beaker in sequence, fully stir and dissolve each mother liquor before the next one is added until all mother liquors are added, and stir to fully dissolve.

## III. Making up to Volume

Transfer the mother liquor mixture into a volumetric flask and rinse the beaker twice or three times. Then transfer the rinsed solution into the volumetric flask and make up to the required volume.

## IV. Boiling

Pour the solution with constant volume into an enamel measuring cup, add agar, and boil until the agar is completely dissolved. Alternatively, omit the boiling step, adjust the pH value after making up to volume, fill directly, and dissolve the agar by autoclaving. During filling, the agar should be evenly filled to avoid hard media or non-coagulated media.

## V. Adjusting the pH

Adjust the pH value of a medium by adding 0.1 mol/L HCl dropwise if the medium is alkaline or adding 0.1 mol/L NaOH dropwise if acidic, and then test the pH value with precision pH test strips or pH meters to keep it within 5.8–6.0. During autoclaving, the pH value will drop by 0.2–0.3 units.

## VI. Filling

Transfer the prepared medium into culture flasks with a test tube while it is hot. An appropriate amount of medium, about 30–35 mL in volume and 1.5–2 cm in thickness should be evenly filled in each 100–150 mL culture flask. The medium should not adhere to the flask mouth to avoid contamination.

The flasks must be immediately sterilized after being filled. If not, they should be stored in a refrigerator and sterilized within 24 hours.

## VII. Marking and Recording

The sealed medium should be marked and placed in an autoclave for sterilization.

## VIII. Autoclaving of Medium

(1) Adding water: Add water to the autoclave until the heating wire is submerged (or until between the high and low water level marks).

(2) Loading into the autoclave: Load the culture flasks and culture instruments into the autoclave.

(3) Sealing: Seal air outlets of the autoclave.

(4) Pressurization: Power on and increase the pressure to 0.05 MPa.

(5) Exhaust: Cut the power supply and open the exhaust valve to discharge cold air until the pressure becomes 0 MPa.

(6) Pressure stabilization: Power on and increase the pressure to 0.1 MPa; keep the pressure within 0.1–0.15 MPa for 20–30 min, and keep the temperature within 121–125°C.

(7) Pressure reduction: After pressure stabilization, power off and cool down naturally until the pressure is 0 MPa.

(8) Taking out from autoclave: Open the autoclave to take out the culture flasks and put them on the bench for cooling. If the medium has not been boiled, gently shake the culture flasks when taking them out to well-mix the agar.

(9) Preservation of medium: Preserve the sterilized medium in a clean and contamination-free environment, and use it as soon as possible, generally within 2 weeks.

**Knowledge Points**

1. Knowledge about Sterilization by Filtration

Some plant growth regulators are not thermostable and cannot be added to the medium with other liquors for autoclaving, so they need to be sterilized by filtration. In terms of filter sterilization, put the autoclaved filter, filter membrane, and containers to receive the filtrate, pipette, or pipettor tip on the clean bench. Then put the prepared growth regulators and antibiotics that need to be sterilized by filtration on the clean bench. Sterilize both hands and install the filter on the clean bench. Add the growth regulators to be filtered into the filter funnel or syringe. Start the decompression filter or push the syringe piston rod to allow the solution to flow through the filter membrane. Add a required amount of filtrate into a solidified medium or a cooled liquid medium immediately with the pipette or pipettor (with a sterilized tip).

2. Knowledge about Medium Preservation

After a medium is autoclaved, it should be used as soon as possible, preferably within 7 days, to prevent components such as IAA and $GA_3$ in the medium from being decomposed. The medium not to be used for the time being should be properly stored and protected from light to ensure that the texture and composition of the medium are not affected.

## Task 3 Pre-treatment and Sterilization of Explants

### I. Collection of Explants

Common explants include shoot tips, shoot segments, leaves, fruits, seeds, and anthers. Explants are usually collected in the morning of a sunny day in spring and summer, and the plant materials that grow vigorously and are genetically stable and not affected by pests or diseases should be selected as explants. The tender parts of a young plant are the most preferable.

**Knowledge Points**

The selection of explants is one of the critical factors in plant tissue culture. Suitable explants can be more easily cultured under *in vitro* conditions. Therefore, it is necessary to select the explants according to the following principles.

1. Strong Regeneration Capacity

Tissues or organs with normal growth on healthy plants are selected as explants as they have a high metabolism and strong regeneration capacity. In addition, for vigorously growing cells or tissues, the higher the degree of differentiation, the weaker their regeneration capacity and the less likely the dedifferentiation. Therefore, plant materials with low differentiation shall be selected as explants where possible. In general, young tissues are better than old ones because of their better morphogenetic capacity. In terms of season, the materials shall be collected where possible in the season when the plants are growing vigorously, because such materials have a high content of endogenous hormone and are conducive to redifferentiation.

2. Readily available and genetically stable for materials

To determine which part to select, it is necessary to consider whether there are abundant sources of culture materials and whether the calluses generated by the explant materials will have undesirable genetic variations and lose the good characteristics of the original variety after dedifferentiation, so as to ensure the quality and quantity. Therefore, materials that are easy to obtain and have little variation shall be selected as explants.

3. Easy Sterilization

To reduce the contamination from explants in plant tissue culture, explant materials carrying a few microbes will be selected where possible. Generally, it is easier to sterilize aboveground plant tissues than underground ones, annual tissues than perennial ones, and young tissues than old ones. Greenhouse materials carry fewer microbes than field materials, and materials germinated in an artificial light incubator can be better sterilized.

4. Explant Size

In the tissue culture of many plant materials, it is found that too-large explant materials are difficult to sterilize thoroughly and easy to cause contamination and waste and too-small explant materials tend to cause the formation of many calluses with a low survival rate. The explants shall not be too small unless used to remove viruses. In terms of the critical size of explants to ensure survival, for shoot tip culture, the explant should be a shoot apical meristem with 1–2 leaf primordia with a size of 0.1–0 5 mm; for leaves and petals, they should be 0.5–1.0 $cm^2$; for shoot segments, they should be 0.5–1.0 cm long.

## II. Pre–treatment of Explants

Remove the unnecessary parts of the explants, cut the remaining parts into materials of a proper size according to the tissue culture requirements, and wash them under running water for 2–4 hours (depending on the type of materials).

## III. Sterilization of Explants

On the clean bench, soak the culture material in 70% alcohol for 30 s, then in 0.1% mercuric chloride solution for 10 min, or in 10% bleaching powder solution for 10–15 min, and keep stirring to well mix the plant material and the sterilizing agent; then rinse it with sterile water 5–6 times for inoculation.

In the process of plant tissue culture, the species of explants, the collection season, the part selected, and the pre-treatment method will all affect the number of microbes carried by explants. The type and concentration of sterilants, as well as the length of sterilization, should be selected depending on the type of explants for surface sterilization. Common sterilants and their effects are shown in the Table 3-4.

**Table 3-4 Common Sterilants and Their Effects**

| Sterilant | Working Concentration(%) | Easy or Difficult Sterilization | Sterilization Length(min) | Sterilization Effect |
|---|---|---|---|---|
| Ethanol | 70–75 | Easy | 0.1–3 | Good |
| Mercury chloride | 0.1–0.2 | Difficult | 2–15 | Best |
| Bleaching powder | Saturated solution | Easy | 5–30 | Very Good |
| Calcium hypochlorite | 9–10 | Easy | 5–30 | Very Good |
| Sodium hypochlorite | 2 | Easy | 5–30 | Very Good |
| Hydrogen peroxide | 10–12 | Easiest | 5–15 | Good |

**Knowledge Points**

Surface disinfectants are harmful to plant tissues. The concentration and treatment time of disinfectants should be selected correctly to reduce tissue death. After surface disinfection, the material must be rinsed with sterile water more than three times to remove residual fungicide. But if alcohol is used for disinfection, rinsing is unnecessary. Sections that have been in contact with disinfectants need to be cut before transferring the material to a sterile substrate. This is because disinfectants will prevent plant cells from absorbing nutrients from the substrate. If the explants are seriously contaminated, rinse them with running water for more than one hour or use corresponding seeds to obtain sterile plantlets and then carry out tissue culture with corresponding parts. Mercury chloride has the best disinfection effects, but is the most harmful to people. After usage, the material should be rinsed with water at least five times, and mercury chloride should be recycled.

After surface sterilization and inoculation of explants, if most of the inoculated materials are contaminated, it indicates that the soaking time in the disinfectant is short. If the inoculated materials are not contaminated, but become yellow with soft tissues, it indicates that the sterilization time is so long that the plant tissues have been destroyed and died. If the inoculation material is not contaminated and grows normally, the sterilization time is considered appropriate.

## Task 4 Aseptic Operation Techniques and Post-inoculation Management

### Knowledge Points

In the inoculation room for tissue culture, sterilization is conducted through fumigation. The strong oxidizing potassium permanganate is used to volatilize formaldehyde in the formalin solution. Since formaldehyde can denature and coagulate microbe protein and dissolve the microbe lipoid, it can kill the bacterial propagules, brood cell fungus, and viruses on the surface of an object and in the air, thus killing the pathogenic microorganisms after a certain period of time.

During fumigation, first close the doors and windows of the inoculation room and then add formaldehyde and potassium permanganate in sequence in a beaker at a ratio of 10 $mL/m^3$ to 5 $g/m^3$; leave the room immediately with the windows closed for 20–30 min of fumigation. Then, open the windows; turn on the UV lamp and fan for 30 min before inoculation to ensure that the inoculation room is thoroughly sterilized.

## I. Preparations before Inoculation

The following steps are generally followed before inoculation.

(1) One day before inoculation, fumigate the inoculation room with formaldehyde and keep the ultraviolet lamp on for sterilization for 30 min.

(2) Put the primary culture medium and inoculation instruments, sterile paper, sterile water, sterilants, and other inoculation supplies on the clean bench for later use. Turn on the fan of the clean bench as well as the UV lamp on the bench 30 min before inoculation.

(3) Wash hands, put on a special lab coat and change to slippers in the buffer room.

(4) Before operating on the bench, wipe hands, especially the nails, with absorbent cotton balls, and then wipe the bench surface.

(5) Clean the inoculation tools with absorbent cotton balls, sterilize the tweezers and scissors thoroughly, and then repeatedly burn their tips for sterilization.

(6) During inoculation, the inoculator must keep his or her hands on the bench and may not talk, walk around, or cough, as these actions can increase microbes while keeping quiet helps to reduce the microbes.

## II. Inoculation

1. The Aseptic Operation of Primary Culture

(1) Put the preliminarily washed and cut explants into a beaker. Place the beaker on the clean bench, disinfect it and rinse it with sterile water several times. Drain the water, take out the explants, and put them on sterilized filter paper.

(2) After the material is dried, hold tweezers in one hand and scissors or scalpel in the other hand to cut the material properly. For example, cut the leaves into 0.5 cm tubes and the stems into segments with 1 node; Strip the micro shoot tip so that it has only 1–2 young leaves. Burn inoculation instruments frequently during inoculation to prevent cross-infection.

(3) Insert or place cut explants onto the medium using burned instruments.

Specific operations are as follows.

First, remove the sealing film, hold and tilt the culture flask or test tube to make its mouth close to the flame of the alcohol burner, and rotate and burn the mouth over the flame for several seconds. Second, gently insert a piece of cut explant into the medium with tweezers. If leaves are taken as explants, directly attach 1–3 pieces (with the leaf back upward) to the medium. There are no requirements as to how the materials should be inoculated, except that the shoot tips and stem segments should be inoculated upright (with the tip upward). It is advisable to inoculate only a small number of materials, i.e. one piece of material in one inoculation flask each time, which can save medium and manpower. Once the culture is contaminated, it can be discarded.

(4) After inoculation, the mouth of the culture flask and the sealing film should be rotated and burned over the flame for several seconds, and then the flask should be sealed with the sealing film tied up. Indicate the medium number, culture material, and inoculation date on the culture flasks and send the flasks to the culture room.

(5) Clean the bench and sterilize it with a UV lamp for 30 min after inoculation work is completed. In the case of continuous inoculation, high-level disinfection is required every 5 days.

2. The Aseptic Operation of Subculture

(1) Put primary culture medium, inoculation instruments, aseptic paper, sterile water, disinfectants, and other inoculation supplies onto the clean bench, and turn on the fan of the clean bench as well as the UV lamp on the bench 30 min before inoculation.

(2) Wash hands, wear a lab uniform and then enter the inoculation room. Wipe hands, the bench surface, and inoculation tools with absorbent cotton balls.

(3) Place the alcohol burner about 30 cm from the edge of the clean bench directly in front of the chest, and note that all subsequent operations should be completed within a 10 cm radius around the alcohol burner. Ignite the alcohol burner, burn the scissors and tweezers from top to bottom, then burn their tips repeatedly. Place the explant flasks to the left of the lamp and the blank medium

flasks to the right; Place the Petri dish for inoculation about 15 cm from the edge of the clean bench directly in front of the chest to facilitate operation.

(4) Open the explant flask and burn the inner and outer walls of its mouth. Plantlets are first transferred to Petri dishes with aseptic paper inside. Hold the culture flask horizontally with the left hand, pinch a piece of cut inoculum material with tweezers in the right hand, and put it into the flask. Gently insert it into the medium. Then burn the flask mouth over the flame of the alcohol burner for a few seconds, and quickly cover the flask with sealing film and tie the mouth up. Frequently wipe (or spray) both hands and the bench with 70% alcohol during operation; repeatedly soak inoculation instruments in 95% alcohol and sterilize them over the flame (or with a dry heat sterilizer) to prevent cross-contamination. After inoculation, burn both the mouth and sealing film of the explant flask when needed. For a skilled operator, suspension inoculation may be adopted, i.e. holding the tip of the test-tube plantlet with tweezers to pull it out of the flask, cutting off the tip, and inserting it into the blank medium.

(5) After inoculation of a flask of stock plant material, wipe both hands with absorbent cotton balls, discard the used Petri dish, and replace it with a new sterile one to prevent cross-infection.

(6) After inoculation of 10 flasks of stock plant material, label them, move them down the clean bench into a plastic basket, and inoculate the next batch.

(7) When all the inoculation work is finished, put out the alcohol burner, clean up the clean bench, put plantlet flasks, and then use Petri dishes, empty explant flasks, inoculation tools, and other inoculation supplies into plastic baskets. Take them out of the inoculation room and clean them.

(8) Transfer them to the culture room for culture after inoculation. Check contamination after 3 days and remove any contaminated culture flasks in time.

## III. Control of Environmental Conditions after Inoculation

The inoculated plant materials are usually cultured in the culture room and grow well under appropriate temperatures, usually (25±2)°C, but vary slightly from culture to culture. Under normal circumstances, the lighting intensity of the culture room is required to be 1,000–6,000 lx. The commonly used lighting intensity is 3,000–4,000 lx, and the photoperiod is 14–16 h of light and 8–10 h of darkness. The humidity is generally 70%–80%.

**Knowledge Points**

1. Sterilization of Inoculation Room by Fumigation

In the inoculation room, add formaldehyde and then potassium permanganate into the beaker at a ratio of 10 mL/$m^3$ to 5 g/$m^3$, and leave quickly. Turn on the UV lamp and fan 30 min before inoculation to ensure that the inoculation room is thoroughly sterilized.

2. Regulation of Environmental Conditions in the Culture Room

(1) Temperature regulation of test-tube plantlets

Temperature is an important factor during tissue culture as it affects the differentiation, proliferation, and organogenesis of explants, as well as the growth and development of tissue culture plantlets. The explants grow and differentiate well at an optimal temperature. The temperature for tissue culture in most cases is controlled within a range of 24–28℃, under which shoots and roots can be formed generally. A temperature lower than 12℃ is not conducive to the growth and differentiation of tissue, and a temperature higher than 35℃ is not conducive to the growth of test-tube plantlets. Therefore, a constant temperature of (25 ± 2)℃ is mostly needed in production. Different types of plants vary in their optimal temperatures. The optimal temperature is 20℃ for lily, 25–27℃ for Chinese rose, 28℃ for tomatoes, and 28℃ for the formation of tobacco buds; Variable temperature conditions (28℃ during daytime and 15℃ at nighttime) for *Jerusalem artichoke* are most suitable for the formation of roots. Leaf development rate and percentage of regenerated lily bulblets are higher at 30℃ than at 25℃.

(2) Lighting regulation of test-tube plantlets

As one of the important conditions in tissue culture, lighting primarily plays a role in inducing organogenesis. Lighting intensity, quality, and photoperiod greatly affect test-tube plantlets. For tissue culture of plants of different species, lighting requirements for organogenesis vary. For example, tobacco and parsley do not require lighting for organogenesis.

Illumination plays an important role in the proliferation of culture cells and the differentiation of organs. According to current research, lighting intensity has a significant effect on the initial division of explants and cells. Generally, plantlets grow strong and robust at higher lighting intensity and exhibit excessive growth at lower lighting intensity.

Light quality has a significant effect on callus induction, tissue proliferation, and organ differentiation. For example, lily bulblets differentiate into calluses after 8 weeks of culture under red light and after a dozen weeks under blue light; after *Gladiolus gandavensis* bulblets are inoculated for 15 days, buds appear and plantlets formed grow vigorously if cultured under blue light while plantlets are slender if cultured under white light. Blue light promotes the formation of tobacco calluses and plantlets, while red light and far-red light do not have such an effect. The effect of light quality on root formation is exactly the opposite of that of buds.

The photoperiod also has a significant effect on plant organogenesis. Test-tube plantlets should be cultured within a certain photoperiod, most commonly 16 hours of lighting and 8 hours of darkness. Studies have shown that organ tissues of short-day cultivars are easy to differentiate under short-day sunlight and produce calluses under long-day sunlight. Sometimes dark culture is needed, especially for plants whose calluses grow better in darkness than in light, such as calluses of safflower and Chinese tallow trees. For the culture of grape shoot segments sensitive to short-day sunlight, roots are formed only under short-day sunlight, while for cultivars not sensitive to sunlight, roots could be formed in any photoperiod. For example, the root segments of long-day chicory could be induced to form flower buds under long-day sunlight.

(3) Humidity regulation of test-tube plantlets

The influence of humidity on test-tube plantlets mainly comes from culture containers and the environment. The humidity in culture containers mainly depends on the medium moisture content and sealing material.

The humidity in the culture containers is generally high, usually up to 100%. With the extension of culture time, the humidity gradually decreases. Since the humidity in the culture containers is mainly affected by the agar content and sealing material, when the environmental humidity is low, the amount of agar should be appropriately reduced to increase the humidity in the culture containers. Otherwise, the medium will be dry and hard, which is not conducive to the inoculation of the explants into the medium, resulting in stunted growth and development. The sealing material directly affects the humidity in the containers: The better the seal, the higher the humidity. However, the sealing material with good airtightness is prone to cause poor gas exchange, leading to stunted growth and development.

The relative humidity of the environment can affect the evaporation of water in the medium. Humidity being too high and too low is unfavorable: Low humidity will cause water loss in the medium and affect the growth of test-tube plantlets, while high humidity contributes to the breeding of undesired microbes, resulting in contamination. The culture room generally requires 70%–80% relative humidity. Humidity can be adjusted with a humidifier or by frequent sprinkling when it is too low and with a dehumidifier when it is too high.

## IV. Common Problems and Preventive Measures in Plant Tissue Culture

### (I) Contamination

1. Contamination Phenomenon

Contamination indicates the failure of plant tissue culture because of the growth of undesired microbes in the medium and on culture materials during the process. The main contaminants in plant tissue culture are bacterial and fungal contamination. Mucinous plaques appear with a foul odor 1–2 days after inoculation in case of bacterial contamination and molds in different colors appear 3–10 days after inoculation in case of fungal contamination.

2. Causes

(1) Explant carrying microbes

This is caused by contaminated plantlets in the original flask and recklessly sterilized explants. Generally, perennial woody plants carry more microbes than annual or biennial herbaceous plants; older plants carry more than younger ones;

field-grown plants carry more than greenhouse-grown ones; plants with mud carry more than clean ones. Plants collected on rainy days are more vulnerable to microbial infection while those collected in strong sunshine at noon carry fewer microbes.

(2) Medium carrying microbes

The effects of culture medium sterilization may be affected by autoclave use and autoclave sterilization temperature, pressure, and duration, as well as the pore size of the filter membrane, sterilization of filter sterilization equipment, and operation in filter sterilization during filter sterilization. In addition, loose caps of culture flasks and long-time storage of medium may cause microbial contamination.

(3) Operation procedures

Contamination may be caused if the inoculation room is not clean, dry, or sealed, or is not frequently sterilized by ultraviolet irradiation, formaldehyde fumigation, or 70% alcohol spray; the clean bench is not sterilized and the fan is not turned on; operators are not skilled or operate against relevant standards, or walk around frequently and talk during operation.

(4) Culture environment

The culture room must be kept clean and airtight: Spray it with 70% alcohol every day to remove microbes and dust, and fumigate it with a small amount of formaldehyde every week. Better effects will be achieved if an air filter is installed. Contamination will be aggravated when the relative humidity of the culture room is too high.

3. Preventive Measures

(1) Sterilization of inoculation room

It usually takes 30 min before inoculation each time to scrub the indoor equipment and bench surface with 20% bromogeramine and then irradiate them with a UV lamp for 20 min. Before use, 70% alcohol can be sprayed to quickly reduce the dust in the air. Contaminated materials (if any) must not be cleaned directly in situ and must be autoclaved before being cleaned. Otherwise, a large number of spores will be dispersed. To strengthen the sterilization of the environment, timely cleaning and sanitizing, and regular fumigation are required to ensure a clean environment.

(2) Standard operation

During tissue culture, it must be guaranteed that the media, utensils, and appliances are thoroughly sterilized. Moreover, strict aseptic operations are required during inoculation. Inoculators must be skilled and competent and must wipe their hands and wrists with absorbent cotton balls at the time of inoculation.

(3) Selection and sterilization of explants

To reduce the occurrence of contamination, explants should be carefully selected as the number of microbes they carry varies with the type, part, size, age, and collection period of explants. Generally, it is advisable to collect healthy and robust plants in spring: At noon on consecutive sunny days, select clean and vigorous parts, and sterilize them on site. Explants can be germinated through liquid culture indoors or under aseptic conditions: Brush the branches with a small amount of washing powder or dilute soap solution for hydroponics or liquid culture. New twigs or buds should be selected as explants where possible to reduce the contamination rate. Alternatively, branches collected from the field may be cultured in darkness under aseptic conditions, and etiolated water sprouts should be taken as explants to effectively reduce contamination. The explants should be thoroughly sterilized multiple times or soaked alternately in various sterilants to reduce the microbes carried.

(4) Disposal of contaminated materials

Materials found to be contaminated must be disposed of promptly; Otherwise, it will cause environmental contamination in the culture room. For some particularly valuable materials that are contaminated, they may be sterilized under more stringent conditions and then re-cultured in a fresh medium. The contaminated culture flasks to be disposed of should be intensively autoclaved before being uncapped to remove contaminants and then washed for later use.

## (II) Browning

1. Browning Phenomena

Browning is a process in which polyphenol oxidases in explants are activated during culture to oxidize phenolic substances in cells into brown quinones. These brown quinones not only spread outwards to gradually turn the medium brown but also inhibit the activity of other enzymes, seriously affecting the dedifferentiation and organ differentiation of explants and eventually leading to browning and perishing of explants.

2. Causes

(1) Plant cultivar

Generally, woody plants have a higher content of phenolic compounds than herbaceous plants, making them more vulnerable to browning during tissue culture.

(2) Physiological status of explants

The ageing degree, age, size, and selected part of the explants are all influencing factors of browning. Higher ageing degree of explants means higher lignin content and greater likelihood of browning. Larger explants are more prone to browning than smaller ones; older materials are generally more severely browned than younger ones. The larger the cut, the more the phenolic substances oxidized and the more severe the browning.

(3) Medium composition

In the case of a high concentration of inorganic salt and a high CTK level in the medium, cells will proliferate and inhibit browning under the most suitable conditions for the dedifferentiation of explants. Browning is less likely to occur in liquid culture but more likely to occur at a high pH level.

(4) Culture conditions

Strong sunlight, high temperature, and long culture time will accelerate browning.

(5) Material transfer time

Test-tube plantlets are prone to browning if left uninoculated for a long time.

3. Preventive Measures

(1) Select suitable explants

Select tender parts of young plants that are appropriately sized and well-grown as explants.

(2) Select suitable culture conditions

Initial culture at moderately low temperatures in darkness or weak light can inhibit the oxidation of phenols and reduce browning. A reduced amount of agar and a lower pH level are conducive to reducing browning and a suitable concentration of cytokinin in the medium can also reduce or inhibit browning.

(3) Speed up subculture transfer

Continuous transfer of the explants to a fresh medium can prevent the accumulation of polyphenol oxidase and reduce browning.

(4) Add antioxidant

Antioxidants such as sodium thiosulfate, phytic acid (PA), ascorbic acid, hydrogen peroxide, and cysteine should be added to the medium. Absorbents include activated carbon (AC) and polyvinylpyrrolidone (PVP). Antioxidants should be used in stages. As some antioxidants may have toxic effects on the culture, long-term culture in the medium containing these antioxidants must be avoided. If browning is controlled in the early stage, the antioxidants should be removed from the medium.

(5) Add activated carbon

Adding 0.1%–0.5% activated carbon to the medium can reduce browning since activated carbon can adsorb polyphenol oxidase. However, activated carbon adsorbs nutrients and hormones as well at the time of adsorbing harmful substances, thus affecting the growth and development of explants.

## (III) Vitrification

1. Vitrification Phenomena

Vitrification is a physiological disorder of test-tube plantlets. During *in vitro* culture, the tender stems or leaves of some cultures appear translucent or watery because of high water content. Vitrified plantlets feature short stalks, narrow internodes, and translucent leaves that are brittle and fragile. In vitrified plantlets, the contents of dry matter, chlorophyll, protein, cellulose, and lignin are low, and no cuticle or functional stoma is found on the leaf surface. Vitrified leaves only have spongy mesophyll (no palisade mesophyll) and are characterized by decreased photosynthetic capacity and enzymatic activity, tissue deformity, organ dysfunction, reduced differentiation ability, difficulty in rooting, and low survival rate after transplanting. Vitrified plantlets are common in plant tissue culture, with a proportion of up to 50% in some cases, seriously affecting the reproduction rate and causeing a great waste of human, financial, and material resources.

2. Causes

(1) Inappropriate hormone concentration

Overly high cytokinin concentrations or an imbalanced ratio of cytokinins to auxins can increase the number of vitrified plantlets.

(2) Low agar and sucrose concentrations

Agar and sucrose concentrations are negatively correlated with the percentage of vitrified plantlets, i.e. the higher the agar or sucrose concentration, the lower the percentage of vitrified plantlets.

(3) Culture conditions

The water content of the medium affects the formation of vitrified plantlets, i.e. the higher the water content in the medium, the more severe the vitrification. Low temperature and high humidity also tend to induce vitrification.

(4) Poor ventilation

Poor ventilation may also result in vitrification.

3. Preventive Measures

(1) Increase agar concentration

During solid culture, increasing agar concentration and purity can improve the hardness of the medium, thus blocking the water absorption of cells and reducing vitrification.

(2) Increase sucrose concentration

Increasing the sucrose concentration of the medium can increase the osmotic pressure, lower the osmotic potential of the medium and reduce the water content of the culture material.

(3) Improve medium

Increase the content of Ca, Mg, Mn, K, P, and Fe elements and decrease the content of N and Cl elements in the medium; decrease the concentration of ammonium nitrogen and increase the content of nitrate nitrogen. Properly reducing cytokinin concentration is an important measure to overcome vitrification.

(4) Improve culture conditions

Seal the flask opening with a porous sealing film or a cap with an air-permeable liner. Increasing natural light, controlling the photoperiod, lowering the temperature, and reducing humidity can effectively control the occurrence of vitrification.

(5) Add other substances

Adding activated carbon, paclobutrazol, phloroglucin, and CCC into the medium can effectively reduce and prevent vitrification.

## Task 5 Design of Medium Formulation

The medium formulation is the core of the whole plant tissue culture system. The medium varies greatly with culture materials, stages, and purposes. Therefore, it is of great importance to design a suitable medium formulation to meet the requirements of rapid plant tissue culture and shorten the culture period.

### I. Determine Culture Materials

Select materials with a certain value for plant tissue culture.

### II. Data Consulting

Consult relevant data of the selected materials, such as literature in relevant journals and doctoral dissertations via CNKI, with a focus on the process and formulation of tissue culture of plants of the same family and genus, determine the minimal medium and the main plant growth regulatory substances in need.

### III. Design Medium Formulation Based on the Information Gathered

1. Minimal Medium

The basic-element solutions commonly used include MS, White, $B_6$, and $N_5$. The main difference between the formulations of each element solution lies in the ratio between nitrate-nitrogen and ammonium-nitrogen. The minimal medium can be adjusted as required if there are other special needs for basic elements.

2. Carbon Source

The carbon source is mainly sugar, and sucrose is commonly used sugar.

The sucrose used in the laboratory is analytically pure and relatively expensive. Generally, edible sucrose can be used as a substitute if production requirements are not high.

3. Medium Coagulant

Common coagulants are agar, carrageenan, and gelatin. However, for liquid medium, coagulants are not needed.

4. Plant Growth Regulator

Plant growth regulators are the most important part of the medium formulation where design varies most significantly. The two main types of plant growth regulators commonly used in plant tissue culture are auxin and cytokinin. There are certainly other regulatory substances with specific effects, such as growth inhibitors. The main considerations in formulation design are the type and amount of addition and the ratio between the two types of regulators.

5. Additive

Common additives include activated charcoal, coconut juice, and bananas, all of which have a certain effect on plant growth.

6. pH Value

The pH value of most media is 5.8, which can be adjusted if specifically required.

To be specific, change one component of the formulation while keeping the rest unchanged in the design of the medium formulation. In addition, a gradient should be set to observe the change of plants and screen for the best medium according to the designed gradient.

**Knowledge Points**

Common Minimal Media

(1) MS medium

MS medium was designed by Murashige and Skoog in 1962 for culturing tobacco cells. It is characterized by a higher concentration of inorganic salt and a higher content of potassium salt, iron salt, and nitrate than other media. There are abundant nutrients in the MS medium in proper quantity and proportion, and it can meet the needs of plant tissue growth without adding more additives. With the effect of accelerating culture growth, it is currently the most widely used medium.

(2) White medium

White medium was designed by White in 1943 for culturing tomato root tips. It is characterized by a low concentration of inorganic salt ions, and is suitable for rooting culture.

(3) $B_5$ medium

$B_5$ medium was designed by Gamborg et al. in 1968 for culturing soybean root cells. It is mainly characterized by a high content of potassium salts and thiamine hydrochloride, as well as a low content of ammonium salt, which may inhibit the growth of many cultures. Dicotyledonous plants, especially woody plants, show better growth in the $B_5$ medium.

(4) $N_6$ medium

$N_6$ medium was designed by Zhu Zhiqing et al. in 1974 for anther culture of cereal crops such as rice. With a simple composition and a high content of $KNO_3$ and $(NH_4)_2SO_4$, it has been widely used in anther culture and tissue culture of wheat, rice, and other plants in China.

(5) Nitsh medium

Nitsh medium was designed by Nitsh in 1951. It is characterized by a low content of major elements, a few types of trace elements, and a high nitrogen content, and is mainly used for anther culture.

(6) Miller medium

Miller medium was designed by Miller in 1963. Compared with MS medium, the content of inorganic elements in this medium is reduced by 1/3–1/2, and there is a reduced variety of trace elements and no inositol. This medium is mainly used for anther culture.

## Summary

1. Preparation of MS Medium Mother Liquor

Determine medium formulation → Determine the type of mother liquor →

Calculate → Weigh → Dissolve → Mix and make up to volume → Fill→ Label → Store in the refrigerator.

2. Preparation of MS Medium

Determine medium formulation and amount → Calculate the amount of mother liquor → Transfer mother liquor → Weigh agar and sucrose → Make up to volume → Prepare medium → Adjust the pH level → Fill → Seal → Mark and record → Autoclave.

3. Pretreatment and sterilization of explants

Collect explants → Cut into suitable sizes and rinse under running water for 2–4 h → Soak in 70% alcohol for 30 s on a clean bench → Soak in 0.1% mercuric chloride solution for 10 min (or soak in 10% bleaching powder solution for 10–15 min) → Rinse with sterile water 5–6 times → Inoculate.

4. Aseptic Operation Techniques and Post-inoculation Management

Pre-inoculation preparation (sterilization of inoculation room by fumigation, putting in inoculation supplies, and turning on UV lamp for 30 min) → Inoculation → Transfer to culture room for cultivation (temperature: 25 ± 2°C, lighting intensity: 3,000–4,000 lx; photoperiod: 16 h of light and 8 h of darkness; humidity: 70%–80%).

5. Common Problems and Preventive Measures in Plant Tissue Culture

(1) Contamination causes: Explant-carrying microbes, medium-carrying microbes, improper operation, and culture-environment-carrying microbes. Preventive measures include regular and thorough sterilization of the inoculation room, standardized operation, selection of suitable explants and sterilization methods, and timely disposal of contaminated materials.

(2) Browning causes: Vulnerability of cultivars to browning, explants being relatively mature, presence of browning-inducing components in the medium, improper culture conditions, and untimely material transfer. Preventive measures include choosing suitable explants and culture conditions, speeding up subculture transfer, adding antioxidants, and adding activated carbon.

(3) Vitrification causes: Inappropriate hormone concentration, low agar and sucrose concentrations, and poor culture and ventilation conditions. Preventive

measures include increasing agar and sucrose concentration, improving medium and culture conditions, and adding other substances.

6. Design of Medium Formulation

Select culture materials → Consult relevant information to determine minimal medium → Adjust growth regulator concentration and test → Determine formulation according to test results.

7. Acclimatization and Transplanting of Tissue Culture Plantlets

Acclimatize plantlets (store them at room temperature for two days and uncap after seven days) → Clean flask plantlets→ Prepare substrate → Transplant plantlets.

8. Management after Transplanting

Spray bactericide once after transplantation, spray water daily to moisturize, irrigate the land after five days, observe and pick out the dead plantlets after 10 days, and add an additional 0.1% urea or 1/2 MS aqueous solution of major elements one month later.

# Review Test

## I. Explain the Glossary

1. Medium Mother Liquor
2. Contamination
3. Browning
4. Acclimatization

## II. Fill in the Blanks

1. The nutrients required for plant tissue culture are mainly obtained from the medium, and the nutrients in most media mainly include _________, _________, _________, _________, _________, and _________.

2. MS mother liquor can be made of three, four, or five types of liquor. Among

them, the one made of four types of liquor including ________, ________, ________, and ________ is the most commonly used.

3. The concentration factor of each type of mother liquor in plant tissue culture varies slightly. Generally, the concentration factor is ________ for major elements, ________ for trace elements, ________ for iron salts, and ________ for organic matter.The concentration of plant growth regulator mother liquor is generally ________, ________ mL for each preparation.

4. Generally, a balance with a sensitivity of ________ is used for weighing major elements and iron salts, and a balance with a sensitivity of ________ is used for weighing trace elements and organic matter.

5. The prepared mother liquor is poured into reagent bottles and labeled. The iron salt shall be sealed in a ________ (color) reagent bottle, marked with ________, ________, ________, and ________, and then stored in a 4°C refrigerator.

6. Plant growth regulators include ________, ________, ________, ________, and ________, and the most commonly used in plant tissue culture are ________ and ________.

7. Commonly used auxins include ________, ________, ________, and ________.

8. Commonly used cytokinins include ________, ________, and ________.

9. After inoculation, flasks are clearly labeled with ________, ________, and ________, and sent to the culture room.

10. The lighting intensity of the culture room is required to be ________ lx, the commonly used lighting intensity is ________ lx, and the photoperiod is ________ h of light and ________ h of darkness.

11. The substrate used in the transplanting of test-tube plantlets is a compound substrate composed of ________, ________, and ________ in the ratio of ________, and scraped flat in the plug trays.

## III. True or False Questions

(　　) 1. In terms of medium autoclaving, the higher the temperatures, the more complete the sterilization.

(　　) 2. For plant tissue culture, the ratio of cytokinin to auxin controls the organ development pattern. A higher concentration of auxin promotes root formation, while a higher concentration of cytokinin promotes bud differentiation.

(　　) 3. In the selection of explants, the higher the degree of differentiation, the easier the dedifferentiation of well-grown cells or tissues.

(　　) 4. For explant inoculation, the fan of the clean bench and the UV lamp on the bench should be turned on 30 min before inoculation, and inoculation should be carried out 30 min after the UV lamp is turned off.

(　　) 5. During plant tissue culture, the pH value of most media is 5.8.

## IV. Multiple Choice Questions

1. The temperature for medium autoclaving is _________.

A.90–100°C　　B.110–120°C　　C.121–126°C　　D.180–200°C

2. Among the commonly used disinfectants, the one with the best disinfection effect is _________.

A.Ethanol　　B. Mercury chloride

C. Bleaching powder　　D. Sodium hypochlorite

3. The temperature of the culture room is usually _________.

A. (25±2)°C　　B. (18±2)°C　　C. (20±2)°C　　D. (30±2)°C

4. The relative humidity in the culture room is generally _________.

A.70%–80%　　B.60%–70%　　C.50%–60%　　D.80%–90%

5. The most thorough way of space sterilization in plant tissue culture is _________.

A. by fumigation　　B. by UV

C. with 75% alcohol　　D. with hydrogen peroxide

## V. Short Answer Questions

1. What are the principles of explant selection for plant tissue culture?

2. What are the common problems in plant tissue culture? How to prevent

them?

3. What are the steps of autoclaving media?

4. What principles should be followed in explant selection for plant tissue culture?

5. What are the common problems and preventive measures in plant tissue culture?

## VI. Calculation

1. Calculate the amount of each chemical weighed and the volume used for the preparation of 1 L medium according to the concentration factor and the preparation volume of mother liquor in the form for MS medium mother liquor preparation, and record the results in the Table 3-5.

**Table 3-5 Preparation Table of MS Medium Mother Liquor**

| Name of Mother Liquor | Component | Prescribed Amount (mg/L) | Concentration Factor | Weighed Amount (mg) | Volume of Mother Liquor (mL) | Volume Pipetted for 1 L Medium (mL) |
|---|---|---|---|---|---|---|
| Major elements | $KNO_3$<br>$NH_4NO_3$<br>$MgSO_4 \cdot 7H_2O$<br>$KH_2PO_4$<br>$CaCl_2 \cdot 2H_2O$ | 1,900<br>1,650<br>370<br>170<br>440 | 20 | | 1,000 | |
| Trace elements | $MnSO_4 \cdot 4H_2O$<br>$ZnSO_4 \cdot 7H_2O$<br>$H_3BO_3$<br>KI<br>$Na_2MoO_4 \cdot 2H_2O$<br>$CuSO_4 \cdot 5H_2O$<br>$CoCl_2 \cdot 6H_2O$ | 22.3<br>8.6<br>6.2<br>0.83<br>0.25<br>0.025<br>0.025 | 100 | | 1,000 | |
| Iron salts | $Na_2EDTA$<br>$FeSO_4 \cdot 4H_2O$ | 37.3<br>27.8 | 50 | | 1,000 | |

continued

| Name of Mother Liquor | Component | Prescribed Amount (mg/L) | Concentration Factor | Weighed Amount (mg) | Volume of Mother Liquor (mL) | Volume Pipetted for 1 L Medium (mL) |
|---|---|---|---|---|---|---|
| Organic matter | Glycine<br>Thiamine hydrochloride ($VB_6$)<br>Pyridoxine hydrochloride ($VB_1$)<br>Nicotinic acid<br>Inositol | 2.0<br>0.1<br><br>0.5<br>0.5<br>100 | 200 | | 1,000 | |

2. Calculate the amount of each chemical weighed and the volume of solvent used in the preparation according to the concentration and preparation volume of mother liquor in the form for preparation of plant growth regulator mother liquor, and record the results in the Table 3-6.

**Table 3-6　Preparation of Growth Regulator Mother Liquor**

| Name of Mother Liquor | Chemical | Dissolution Medium | Mother Liquor Concentration (mg/mL) | Volume of Mother Liquor (mL) | Weighed Amount (mg) |
|---|---|---|---|---|---|
| NAA mother liquor | NAA | | 0.1 | 200 | |
| 6-BA mother liquor | 6-BA | | 0.2 | 100 | |

# Module 4 Tissue Culture of Virus-free Plantlets

## Work Tasks

### Task 1 Virus-free Strawberry Plantlets by Shoot Tip Culture

Plant viral disease is the second severest disease after fungal disease. For plants infected with viruses, physiological metabolism will decrease with symptoms such as mottled leaves, etiolation, crinkling, and spots, which seriously affect photosynthesis and other physiological functions of plants, inhibit plant growth, and cause morphological distortion, reduced yield, and deteriorated quality. Especially for crops of asexual propagation, such as potatoes, garlic, apples, grapes, strawberries, and flowers, viruses can be transmitted through vegetative organs and cause increasingly serious harm as they accumulate in the mother plant from generation to generation.

Viruses may cause huge damage to plant production. For example, potato virus disease remarkably reduces the yield by 10%–20% annually; grapevine fanleaf virus reduces the yield by 10%–18%; strawberry virus disease severely reduces yield and quality; flower virus generally affects the ornamental value, resulting in deformed, discolored, and fewer and smaller flowers.

Given the above, countries around the world have begun to lay emphasis on research in this area. However, agents for effective control of virus diseases are still lacking. Although pesticides can control bacterial and fungal diseases, they do not work for virus diseases. Some chemicals, such as thiouracil and nucleic acid preparations, can inhibit virus replication but will also toxicate the host plants. The use of certain plant growth promoters and additional fertilizers can only temporarily alleviate or conceal the symptoms of virus diseases, and cannot solve the problem fundamentally, bringing higher production costs. Heat treatment can inactivate certain viruses but does not affect filoviruses or baculoviruses, which mainly harm plants. Tissue culture is currently the most effective method to prevent and control virus diseases, as it provides virus-free plants by removing the virus from vegetative organs for further expanded propagation in production.

In addition to viruses, this technique can also remove a variety of fungal, bacterial, and nematode diseases to restore seed characteristics and promote plant growth. As a result, less fertilizer and pesticide is needed, and stress resistance is enhanced, lowering production costs, promoting environmental protection, and forming a benign ecological cycle.

## I. Medium Preparation

(1) Primary medium: MS + 0.5 mg/L 6-BA + 3% sucrose + 0.7% agar.

(2) Subculture medium: MS + 0.5 mg/L 6-BA + 3% sucrose + 0.7% agar.

(3) Rooting medium: 1/2 MS + 0.5 mg/L NAA + 3% sucrose + 0.7% agar.

**Knowledge Points**

A suitable medium is conducive to obtaining complete plants through shoot tip culture. Generally, White medium and MS medium are used as the minimal medium. Increasing the content of potassium and ammonium salts will boost the growth of shoot tips in particular. Large shoot tips can form complete plants in a hormone-free medium, but adding 0.1–0.5 mg/L of auxin or cytokinin or both is often conducive to that. 2,4-D can induce callus formation in explants and should be avoided.

Solid media are generally used in shoot tip culture. However, liquid media can also be used in cases where solid media can induce callus formation in explants. For liquid culture, a filter paper bridge must be made. Both arms of the bridge should be immersed into the medium in the test tube, and the bridge deck should be suspended over the medium, with explants placed on it.

## II. Selection and Sterilization of Explants

Apical buds can be taken as explants from robust mother plants or stolons. Alternatively, 3-cm-long stolon segments that are robust, newly germinated, and do not land on the ground can be selected as explants from a pest-free field after 3–4 sunny days from July to August when stolons are at their growth peak.

For sterilization, firstly, rinse with tap water for 2–4 h, wash off the oil on the material's surface with the detergent aqueous solution, and then strip the outer leaves. Secondly, soak in 70% alcohol under sterile conditions for a few seconds to remove the wax from the surface. Finally, sterilize the surface for 15–30 min, keep stirring to promote the permeation of the solution, and then rinse with sterile water 3–5 times.

**Knowledge Points**

The first step for shoot tip culture is to select appropriate explants. To reduce the number of natural microbes on the surface of explants, the test plants can be cultivated in sterile potting soil and transferred to a greenhouse for cultivation before the shoot tips are selected. Corresponding measures should be taken for protected cultivation. For example, water should be directly poured into the soil rather than on the leaves, and the plants should be regularly sprayed with carbendazim or other systemic bactericides. For some field-grown plants, cuttings can be made and cultured in a nutrient solution. The branches grown from axillary buds are much less contaminated than those directly taken from the field.

Well protected by overlapping leaf primordia, the shoot apical meristem is sterile. However, for the sake of safety, surface disinfection of stem buds is still generally required before explants are cut. Buds well wrapped by leaves, such as chrysanthemums, orchids, and ginger, only need to be dipped in 75% alcohol, while buds loosely wrapped by leaves, such as carnation, garlic, and potatoes, should be surface-disinfected with 0.1% sodium hypochlorite for 10 min. In garlic shoot tip culture, bulblets may be dipped in 75% alcohol and then dried to remove alcohol with an alcohol burner.

## III. Primary Culture

Place the culture dish on the stage plate of a dissecting microscope disinfected in advance, hold the shoot tip with fine-point tweezers in one hand, and strip the blades and peripheral primordia layer by layer with a dissecting needle in the other hand. Cut off the meristem with 1–2 leaf primordia with a scalpel when the apical meristem in the shape of a bright semicircle is fully exposed, and inoculate the shoot tip on the medium with the tip pointing upward (one shoot tip for each inoculation container). The shoot tip should be stripped as fast as possible to shorten its exposure time and prevent it from losing water and desiccating. During inoculation, hold the culture flask in the left hand, burn the flask opening and sealing material over a flame, and uncap with the thumb and little finger of the right hand. When the flask is uncapped, make sure that the opening obliquely faces the flame of the alcohol burner to avoid dust falling into the flask and causing contamination. The operator should wipe his/her hands and the bench surface with 70% alcohol frequently during operation and sterilize inoculation tools to avoid cross-contamination.

The shoot tip should be inoculated onto the surface of the pre-prepared primary culture medium for culture at 22–25°C with sunlight exposure of 16–18 h/d and lighting intensity of 3,000 lx. After 2–3 months of culture, the shoot tip can differentiate into bud clusters. Generally, a cluster containing 20–30 small buds is considered suitable. It should be noted that the shoot tip may go dormant under low temperatures and short daylight exposure, so a higher temperature and sufficient photoperiod must be ensured.

**Knowledge Points**

The shoot tip is the primary meristem at the apex of the plant (Fig. 4-1) and exhibits vigorous cell division and high vitality. Viruses are not uniformly distributed in an infected plant. The number of viruses varies with the plant part and age, and the content of viruses in mature tissues and organs is high, i.e. the closer to the shoot tip, the lower the virus content. In the area of about 0.1–1 mm around the growing point, the virus content is almost zero or extremely low. The reason for low virus content in the meristem may be that there is no vascular bundle in the meristem, rendering the virus to be transmitted only through plasmodesma at a rate far lower than the cell division and growth rate. High metabolic activity and endogenous auxin content in vigorously dividing cells also inhibit virus proliferation. Thanks to its good virus elimination effects and stable progeny, shoot tip tissue culture is the most widely used and the most important way to breed virus-free plantlets (Fig. 4-2).

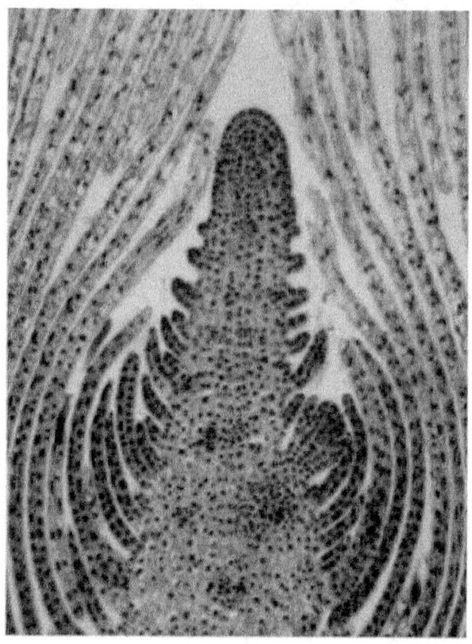

**Fig. 4-1 Structure of Shoot Tip**

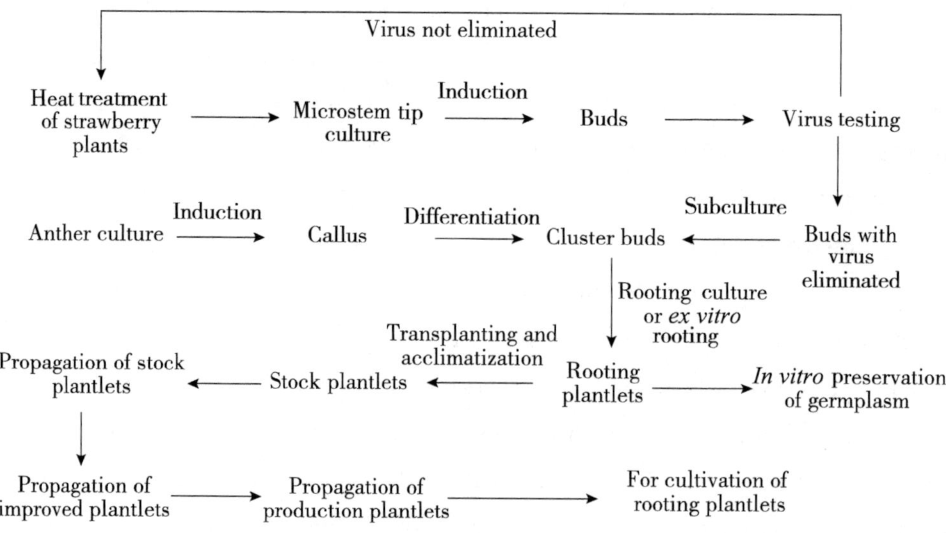


**Fig. 4-2 Cultivation of Virus-free Plantlets Through Virus-free Shoot Tip Culture**

## IV. Subculture

Cut bud clusters into small blocks and transfer them to the subculture medium for the division of plantlets. When the plantlets grow to 1–2 cm, further cut those blocks into clusters with 3–4 buds and then transfer them into a subculture medium for expanding propagation. In terms of incubation conditions, the temperature should be 23–27°C, the lighting intensity should be 1,500 lx, and the photoperiod should be 12 h/d. After 3–4 weeks of culture, bud clusters of 30–40 axillary buds and plants can be obtained.

## V. Rooting Culture

Bud clusters are cut into individual buds with a scalpel and transferred to the rooting medium for rooting. The incubation temperature is 23–27°C, the lighting intensity is 2,000–2,500 lx, and the photoperiod is 12 h/d. After 4 weeks of culture, the bud can grow to robust 4–5 cm high plantlets with 5–6 roots.

**Knowledge Points**

The shoot tip may show 4 main types of growing conditions after inoculation. ① Normal growth: The growing point extends while the leaf primordia develops, basically forming no callus. Buds form in 1–3 weeks and grow into plantlets in 4–6 weeks. ② Growth arrest: The inoculum does not expand and gradually loses vigor until withering. This is mostly because of injury to the shoot tip during stripping. ③ Slow growth: Although the shoot tip is alive and gradually turns green, it increases slowly in size, does not extend, and looks like a green dot, indicating that the culture conditions are not suitable. It is advisable to adjust the medium quickly, transfer the shoot tip to the medium with NAA higher than 0.1 mg/L, and increase the culture temperature appropriately. ④ Overgrowth: After inoculation, the shoot tip growing point does not extend or slightly extend, with a large number of loose translucent calluses forming at the base of the shoot tip. This necessitates timely transfer to an auxin-free medium under a reduced culture temperature, which can inhibit callus growth and promote its differentiation.

## Task 2 Virus–free Strawberry Plantlets by Anther Culture

**Knowledge Points**

The anther culture method is to collect late-uninucleate-stage strawberry buds that have grown to 4–6 mm in size after emergence, strip and culture the anthers under sterile conditions to induce the formation of callus and then adventitious buds, and finally differentiate into independent individuals with stems and leaves. The advantage of anther culture is that the viruses can be effectively eliminated during the process from callus formation to differentiation into stems and leaves. This method allows the cultivation of virus-free plantlets where the virus species are unclear, and there is a lack of indicator plants for virus detection.

When obtaining octoploid-regenerated plants by anther culture, Katsuji Oosawa found that the regenerated plants were virus-free with a 100% elimination rate. In China, Qin Lanying et al. were the first to induce regenerated plants through anther callus and have cultivated virus-free plantlets of Harunoka, Hokowase, Dana, and more cultivars with this method. Xue Guangrong et al. also cultivated virus-free plantlets of 5 cultivars, such as Surprise des haues by direct differentiation of anther callus to plants. Most scholars believe that the virus elimination rate of anther culture is 100%.

### I. Medium Preparation

Callus-inducing medium: MS + 0.5 mg/L 6-BA+ 0.2 mg/L NAA.

Bud differential medium: MS + 0.5 mg/L 6-BA+ 0.5 mg/L KT.

Proliferation medium: MS + 0.5 mg/L 6-BA + 0.1 mg/L GA.

Rooting medium: 1/2 MS + 0.1–0.2 mg/L IBA.

### II. Material Selection

Anthers of the late uninucleate stage are preferable in strawberry anther culture. The method of checking the pollen development period is to collect several flower buds from the field, put 1–2 anthers from each bud on the slide, add 1–2 drops of magenta acetate, crush anthers with a glass rod or tweezers, remove the residues, and cover with a coverslip for microscopic examination. Observe several fields of view. If most pollen has only one nucleus and is squeezed to one side, it indicates that the corresponding anther is in the late uninucleate stage and can be used as an explant. Usually, strawberry buds with pollen in the late uninucleate stage are unopened; the calyx is slightly longer than the corolla; the corolla just shows white, or becomes white or light green, and is not loose, and the anther is yellowish and full.

**Knowledge Points**

Whether the pollen is in a proper development stage is the key to the success of anther culture. Pollen in the mid-to-late uninucleate stage is most likely to form pollen embryos or calluses. For different plant species, microspores in a proper development stage are required to induce callus and embryoid.

The early third stage of microspore development (microspores undergo secondary mitosis to form mature pollen) is the critical stage of embryo formation. For *in vitro* culture, the embryoid cannot be formed if the microspores go beyond this stage of development. Niesch (1968) proposed that during the binucleate stage of the microspore, starch gradually accumulates, preventing the microspore from regenerating into plantlets.

During microspore development, the balance of endogenous hormones in anthers changes continuously. As pollen matures, the hormone balance becomes unsuitable for microspore dedifferentiation, or as pollen matures, the essential substances for pollen dedifferentiation are depleted, preventing the microspore from regenerating into plantlets.

Before anther culture, a microscopic examination at the cellular level is required to determine the stage of pollen development.

## III. Material Pretreatment and Sterilization

The flower buds collected from the field should be wrapped with wet gauze and placed in a plastic bag, tied, and stored in a refrigerator at about 4°C for 24–48 h treatment in darkness.

Wash the inoculation material with tap water and gently dry it with gauze. Then transfer the material into a disinfected wide-mouth bottle and add 70% alcohol to soak it for 30 s; pour out the alcohol and add 3% sodium hypochlorite to disinfect the material for 20–30 min; pour out the disinfectant and rinse the material with sterile water 4 times.

### IV. Inoculation

Inoculate the anthers onto the callus-inducing medium using sterilized tweezers or inoculating loops. High inoculation density is preferred.

### V. Culture

After inoculation, place the material in the culture room for incubation at 25°C with a lighting intensity of 1,000–3,000 lx and a photoperiod of 12 h/d. After 15–20 d of culture, the anthers will begin to form a callus, which can be transferred to the bud differential medium after growing to 2–4 mm.

### VI. Rooting and Doubling

Green plantlets and adventitious roots do not emerge at the same time. When the green plantlets expand to a certain number and reach 3 cm high, they should be transferred to a proliferation medium for proliferation culture or a rooting medium to induce rooting.

The natural doubling rate through strawberry anther culture is high, especially for subculture materials. Therefore, artificial doubling treatment is not required.

## Task 3 Virus Elimination of Shoot Tip by Heat Treatment

### I. Medium Preparation

Shoot tip medium: MS + 0.2 mg/L GA3 + 0.5 mg/L 6-BA + 0.05 mg/L NAA.

Subculture medium: Conventional MS medium.

### II. Material Selection

Before the culture of shoot apical meristems, select from the field potato plants and tubers of the cultivar for virus elimination and rejuvenation during the development stage: An individual plant (or asexual line) that grows vigorously with typical characteristics of the cultivar should be selected together with an individual plant with high yield, high rate of large potatoes and no disease spots as the base material for shoot tip virus elimination to improve the virus elimination effects.

## III. Heat Treatment

To improve the virus elimination effects, the material for virus elimination should be heat-treated before the shoot tip tissue is stripped to inactivate and eliminate the potato leaf curl virus. The selected potato tubers are subject to dormancy breaking and germination treatment. When the apical buds of the potato tubers grow to 1 cm, they are transferred to a light incubator for 12 h illumination per day at 3,000 lx and 37°C for 6–8 weeks.

**Knowledge Points**

The main reason why heat treatment can remove viruses is that when plant tissues are exposed to a certain range of temperatures higher than normal, the viruses inside them will be partially or completely inactivated by heat, but the host plant tissues will be seldom or not harmed. This is because viruses and plant cells show different tolerance to high temperatures. High temperatures can delay the spread of viruses and inhibit their proliferation, thus reducing the virus content continuously. After heat treatment for a period of time, the viruses will be eliminated. There are several common heat treatment methods.

1. Hot water immersion treatment

The hot water immersion treatment is to soak the cut scion or planted material in warm water at about 50℃ for several minutes to several hours. This method is simple, easy, and suitable for sugarcane, woody plants, and dormant organs. However, it may easily harm the material and deactivate most plants at 55℃.

2. Hot air treatment

The hot air treatment is to transfer the growing potted plants into a heat treatment room or light incubator where the temperature is maintained at 35–40℃ and the lighting intensity is 1,000–3,000 lx. The length of treatment varies from plant to plant and can be as short as tens of minutes and as long as several months.

For example, the viruses contained in the shoot tips of carnation can be removed by treating them at 38℃ for 2 months, while virus-free potato plantlets can be obtained after being treated at 35℃ for several months. During the treatment process, water should be supplemented in time to prevent dehydration from affecting the normal metabolic activities of the plantlets. To reduce the damage of high temperature to plantlets, variable temperature treatment may be adopted to improve the survival rate and virus elimination rate.

Virus elimination by heat treatment also has some limitations. On the one hand, not all viruses are sensitive to heat treatment. Heat treatment can only partially reduce the content of most viruses in plantlets, making it difficult to obtain virus-free plantlets through mere heat treatment. For example, using this technique can only eliminate the potato leaf roll virus. On the other hand, prolonged heat treatment will cause metabolic disorders in the plant, lower the survival rate, and increase the possibility of cultivar variation. Therefore, this technique should be applied in combination with other methods to obtain good results.

## IV. Material Collection and Sterilization

Cut several 2–3 cm long buds from the germinated tubers after heat treatment. Gently brush them one by one with a soft brush, place them in a beaker, seal the beaker with gauze, and rinse it with tap water for half an hour. Then, sterilize the buds strictly on the clean bench: Dip them in 75% alcohol for 15 s, and rinse them twice with sterile water. Subsequently, soak in 3% sodium hypochlorite for 15–20 min, rinse 5–8 times with sterile water, 3–5 min each time (keep shaking the container holding the material during rinsing to ensure a more thorough rinsing), and place them in sterile culture flasks for later use.

## V. Shoot Tip Stripping and Inoculation

On the clean bench, place sterilized buds under a 40x dissecting microscope. Hold the buds with tweezers in one hand, and carefully strip the leaves layer by layer with a sterilized dissecting needle in the other hand until the bright round growing point is exposed. Carefully cut the shoot tip with 1–2 leaf primordia and below 0.3 mm in length with a sharp sterile dissecting needle, and then inoculate the shoot tip on the prepared culture medium with the cut touching the agar. Care should be taken to ensure that the cut shoot tip is not in contact with the stripped part, dissecting microscope bench, or tweezers holding the buds. In addition, during the stripping process, the exposure time should be as short as possible, as both the airflow from the clean table and the heat from the alcohol burner will dry out the shoot tip quickly. Therefore, it is better to select a cold light lamp as the lamp source of the dissecting microscope. Also, a piece of sterile wet filter paper should be placed under the material, in this case, each shoot tip, to keep the tip moisturized.

**Knowledge Points**

Plants tend to be infected by multiple viruses rather than a single type of virus. To eliminate different viruses in different plants or the same plant, shoot tips of different sizes are required. Usually, the effects of virus elimination through shoot tip culture are negatively correlated with the size of the shoot tip. For example, in the experiment of removing mottle virus from carnation plants, 6 sizes of shoot tips, namely 0.1 mm, 0.25 mm, 0.5 mm, 0.75 mm, 1.0 mm, and over 1.0 mm, are cut and cultured, and the virus elimination rates are identified to be 66%, 40%, 13%, 11%, 0%, and 0%, respectively, which indicates that the smaller the shoot tip, the better the virus elimination effects. However, the survival rate of cultured shoot tips is negatively correlated with their sizes. The smaller the shoot tip, the more difficult it is to maintain nutrients and water in the shoot tip for a long time, resulting in a reduced success rate and higher requirements for stripping techniques. In practice, both the virus elimination effects and the survival rate should be considered. Generally, 0.2–0.5-mm-long shoot tips with 1–2 leaf primordia are taken as culture material. The presence of leaf primordia also affects the ability of meristem to form plantlets. It is generally believed that leaf primordia provide auxin and cytokinins necessary for the growth and differentiation of meristem.

Due to the limited number of buds sprouting from the tubers and the inevitable damage to the shoot tips resulting from stripping, there are only a small number of strippable shoot tips and a low probability of plantlet emergence, which can easily result in the loss of superior cultivar materials. The following measures may be taken: Collect strictly sterilized tuber buds (1 cm), inoculate them on a hormone-free medium for rapid potato propagation, and culture them at 25℃ under 12 hours illumination per day in the culture room for 3–4 weeks. Cut them into segments when the buds grow into plantlets with 4–5 leaves, transfer them to the same hormone-free MS medium to cultivate plantlets. Propagate for 1–2 cycles and strip the shoot tips once they reach a certain number, which not only avoids the loss of materials but also increases the number of shoot tips and plantlet emergence probability.

## VI. Shoot Tip Culture

Inoculate 2–3 shoot tips per flask and evenly distribute them on the culture surface. Place the inoculated shoot tips in the culture room at 20°C, with a lighting intensity of 2,000–3,000 lx, and 12 hours of light exposure per day.

The growing points of the shoot tip in the tissue culture flasks turn apparently green after 2 weeks of culture; Distinctly elongated stemlets can be observed after 30–40 days, and leaflets can be observed after the formation of leaf primordia. At this time, it can be transferred into an MS medium free of hormones to allow the plantlet to continue to grow and form a root system. After 2 months, it can develop into a plantlet with 3–4 leaves, which will be cut by single nodes for propagation and used for virus detection after surviving.

**Knowledge Points**

For virus elimination techniques in plants, no matter which technique is used to obtain virus-free plants, the plants must eventually undergo strict identification to verify that they are indeed virus-free and are truly virus-free plantlets before they are used in production. Common identification methods include the direct observation method, indicator plant method, antiserum identification method, nucleic acid test, and electron microscopy detection method. The following are some simple and easy-to-use detection methods.

1. Direct Observation Method

The presence of a virus can be determined by directly observing whether the plant grows abnormally and whether visible symptoms caused by a particular virus appear on its stem, leaf, or bud. The leaves of the virus-free plantlets are dark green in color, uniform and consistent in shape, and the plantlets grow well. The plant with viral disease symptoms may be preliminarily identified as a diseased or infected plant based on its weak growth, the presence of chlorotic streak or mottle and curling of the leaves, and dwarfing. The method is easy, intuitive, and accurate. However, the detection is slow and time-consuming as visible symptoms may take an extended period to manifest on the host.

2. Antiserum Identification Method

Plant virus is a nucleoprotein composed of protein and nucleic acid; therefore, it is a preferred antigen that will promote antibody production in animals after injection. The antibody produced in the serum is called antiserum, and the agglutination or precipitation reaction caused by the interaction of antibody and antigen is called serum reaction. Antisera produced by different viruses have their own specificity, and species of unknown viruses can be identified with known antiserum. This antiserum is a highly specific reagent. This method has a high specificity and rapid determination speed and can be completed in a few hours or even minutes. Therefore, the antiserum identification method is one of the most useful methods in plant virus identification.

The common antiserum detection methods mainly include enzyme-linked immunosorbent assay (ELISA), which has the advantages of high sensitivity, high specificity, safety, rapidity, and easy observation. The principle of the ELISA method is to combine the immune reaction of antigen and antibody with the high-efficiency catalysis of enzymes to form an enzyme-labeling immunocomplex. When encountering a corresponding zymolyte, enzymes bound to the complex will catalyze the hydrolysis of the colorless zymolyte to form a colored product, thus allowing the results to be judged qualitatively and quantitatively by visual observation or colorimetry. The ELISA method is easy to operate, requires no special instruments and equipment, and allows easy judgment of results and simultaneous testing of a large number of samples. It has been widely used in the detection of plant viruses in recent years and greatly facilitates the detection, and effective kits are available in the market.

3. Nucleic Acid Test

Direct field observation is simple, easy, and intuitive for the detection of virus-free plantlets. However, if the virus-free plantlets still contain a certain amount of virus and the phenotypic symptoms are not distinct after virus elimination, the virus will accumulate quickly after several generations of propagation. Although the antiserum test can better solve the problem in virus bioassay, this process is complicated by extracting plant virus antigens, injecting them into animals to stimulate antibody production, and then obtaining specific antiserum and finally performing the test, and there are many disturbance factors for plants and animals, which may easily affect the test results. The nucleic acid test is an effective method for plant virus detection. It mainly includes double-stranded nucleic acid electrophoresis, nucleic acid hybridization, polymerase chain reaction (PCR), and gene chip, among which PCR is more commonly used.

PCR is a molecular biological method for rapid amplification of genes *in vitro* or DNA sequences that uses a specific DNA fragment provided as a template and simulates natural DNA replication under the interaction of oligonucleotide primers and DNA polymerase. At present, many plant viruses have been prepared as cDNA marks. The presence of a virus can be detected by extracting the virus from a plant and comparing it with the cDNA of a known virus using RT-PCR during the detection. PCR is highly sensitive with a sensitivity level of 10−15 g. In addition, it shows high specificity and resistance against the interference of heterogeneous proteins in plants.

4. Electron Microscopy Detection Method

The use of electron microscopy allows direct observation of viruses, detecting the presence of viruses, ascertaining the size, shape, and structure of virosome, and identifying the species of virus. This is a relatively advanced method and requires certain equipment and techniques. Recent developments have contributed to the detection of viruses by combining electron microscopy with serology, which is called immunosorbent electron microscopy (ISEM). This method has the advantages of high sensitivity and the ability to determine viruses in plant asperity extracting solution quantitatively.

### VII. Subculture

The identified virus-free potato plantlets can be propagated by solid culture. During the solid culture, take a single segment of the virus-free plantlet and transplant it into a solid medium. Each flask can hold about 20 stem segments. After about 20 days, they can develop into 5–10 cm high plantlets, which can be cut for further propagation. This rapid propagation method allows the plantlet to propagate 5–8 times every month.

## Task 4 Acclimatization and Transplanting of Virus-free Plantlets

Acclimatize rooting plantlets in the test tube for a period of time to allow them to gradually adapt to the external environment before transplanting into a loose and air-permeable substrate. Then, strengthen regulation and pay attention to temperature, humidity, illumination regulation, and timely control of plant disease to improve the survival rate of transplanted plantlets.

The implementation period of this task is long, so the second classroom or production practice should be given priority.

### I. Acclimatization

Move the rooted virus-free plantlets to the greenhouse without uncapping to allow the virus-free plantlets to gradually adapt to the external environment under sunlight and changing temperatures. However, over-temperature in the culture flask should be prevented. If the temperature exceeds 30°C, cooling measures should be taken. After 3–4 days, remove the cap and allow the virus-free plantlets to withstand low temperatures and changes in natural humidity, and such a process lasts only 2–3 days.

### II. Plantlet Lifting and Washing

Pull up the plantlets directly from the culture flask, and then wash off the medium with clean water. In the case of a small opening of the culture flask, tap the culture flask, tilt to move the medium with plantlets to the opening, and then take out the plantlets or dip them into the flask with tools such as bamboo skewers and tweezers to take out the plantlets. The roots and tender buds shall not be damaged during plantlet collection and washing.

## III. Graded Disinfection

The plantlets taken out are divided into rooted/rootless plantlets and strong/weak plantlets. Weak plantlets that are difficult to survive can be discarded directly, and strong plantlets that are not rooted can adopt the cutting method to promote rooting.

Soak the well-graded virus-free plantlets in 1,000-fold dilution of carbendazim, chlorothalonil, or thiophanate-methyl for 2–3 min, drain off the water, and then preserve moisture for later use.

## IV. Transplanting

Spray the prepared substrate with an appropriate amount of water (the humidity is proper when the substrate can be squeezed into a ball when held tightly and scattered when released), load it into a suitable plug tray, and place it on the seedbed, or spread the substrate for transplanting on the seedbed. The washed virus-free plantlets are planted in the substrate and the roots stretch naturally to a depth where the leaves do not touch the substrate. Gently compact the substrate after planting, and try not to damage the roots and leaves.

Soak the rootless virus-free plantlets in a rooting agent before planting. After planting, sprayed with 1,000-fold dilution of chlorothalonil, carbendazim, and thiophanate-methyl for final singling. Covered with film to keep the humidity above 90% and covered with sunshade nets on sunny days.

## V. Post-transplanting Regulation

Post-transplanting regulation of virus-free plantlets after transplanting is critical and includes the following five aspects: temperature, humidity, illumination, nutrition, and pest and disease prevention.

1. Temperature Control

Keep the greenhouse temperature at 18–25°C for 1 week after transplanting, and reduce the temperature difference between day and night. Adjust the temperature in the greenhouse at any time according to the growth period after performing normal management. In summer, the temperature can be lowered by devices like a mist sprayer, fan, and cooling pad; In winter, the temperature in the greenhouse can be regulated by covering multi-layer thermal insulation films and installing heating devices.

2. Maintain Humidity

The air relative humidity shall be maintained at 90%–100% by spraying or covering plastic film in a small plastic tunnel for the first 1 week after transplanting. After a week, ventilate both ends with plastic film and gradually increase the ventilation until the plastic film is removed. Then, keep the leaves moist and robust. After half a month, enter normal regulation and gradually reduce the humidity to make it close to that of the natural environment.

3. Illumination Regulation

Direct sunlight shall not be allowed and shading shall be applied in the early stage of virus-free plantlet transplanting. The small plastic tunnel is used in greenhouses, and covered with a sunshade net. The lighting intensity and photoperiod shall be adjusted according to the recovery, growth, and preference for light or shade of plantlets. Inadequate illumination will affect photosynthesis and cause excessive growth of plantlets; Strong illumination will destroy chlorophyll, which may cause leaf chlorosis, yellowing, or whiting, and stimulate transpiration to escalate water balance. The general lighting intensity is about 1,500–4,000 1x. In addition, we should also pay attention to the time of shading and remove the sunshade net in the morning and evening when the illumination is weak and the temperature is low.

4. Supplement Nutrition

Usually, it takes 20–30 days to enter into normal regulation after transplanting. At this time, the plant's leaves and branches grow upright and robust; the root system grows again, and new roots and tender buds form; the plant does not wilt easily after ventilation, and fertilizer is needed to provide nutrients for its growth. It should be noted to adopt proper concentration and amount in the first application of fertilizer and apply

0.15%–0.20% nutrient solution according to the plant's demand for nutrients. In actual operation, 1,500–2,000-fold dilution of water-soluble fertilizer with the ratio of N, P, and K at 20 : 10 : 20 or 14 : 0 : 14 is generally applied alternately to the plantlets at the plantlet stage every 10–15 days. The principle of "more fertilizer for large plantlets, while less for small plantlets; more fertilizer for strong plantlets, while less for weak plantlets" should be followed. In the middle and late stages of plantlet growth, water-soluble compound fertilizer with the ratio of N, P, and K at 15 : 15 : 15 and foliar fertilizers like potassium dihydrogen phosphate can be applied. Calcium, Ferrum, and trace elements can also be applied to some varieties.

5. Pest Control

The strawberry virus is mainly transmitted by aphids, which mainly include strawberry stem aphids, green peach aphids, and cotton aphids. The most widely distributed among them is the strawberry stem aphid, which features spike-like hairs on its body and can parasitize all parts of the strawberry. The strawberry virus is transmitted by aphids sucking sap in a short time. Contact pesticides such as malathion emulsion and omethoate emulsion can be used for its prevention and control from May to June and September to October. Especially, applications from September to October can prevent aphids from overwintering.

To ensure that the seedlings are virus-free, the production of the breeder seedlings should be carried out in the isolation net room, which is better equipped with a No. 300 insect-proof screen.

## Summary

1. Plant Virus Elimination Methods

(1) Virus Elimination Method by Shoot Tip Culture: Explant selection → Explant sterilization → Shoot tip stripping → Differentiation culture → Proliferation culture → Acclimatization and transplanting.

(2) Virus Elimination Method by Anther Culture: Anther selection → Explant sterilization → Differentiation culture → Proliferation culture → Chromosome

doubling → Acclimatization and transplanting.

2. Significance of Plant Virus Elimination

Restore the original good characteristics of the plant, enhance growth vigor, increase yield, and improve quality.

3. Identification of Virus-free Plantlets

Direct detection, antiserum identification, nucleic acid test, electron microscope identification, etc.

# Review Test

## I. Explain the Glossary

1. Virus-free Plantlets
2. Virus Elimination by Heat Treatment
3. Direct Detection
4. Serum Reaction
5. Germplasm Resources
6. Virus-free Original Seed

## II. Fill in the Blanks

1. In general, the viral content is _____ in mature tissues and organs and _____ in immature tissues and organs, and the growing point (_____ mm region) contains little or no virus.

2. The apical meristem refers to the part above the tenderest _____ of the stem, and the shoot tip is composed of _____ and young leaf primordia below it.

3. Generally, the higher the concentration of virus in the host, the normal protein content in the host, and the _____ the treatment time, the greater the required heat for passivation, that is, the higher the temperature.

4. The main detection techniques used at home and abroad include direct detection, _____, antiserum identification, _____ and _____ identification.

5. In the case of the virus elimination method by microstem tip culture, the size of the explants should be determined by a combination of plantlet rate and virus elimination rate and is generally _____ mm with _____ leaf primordia.

6. In the case of the virus elimination method by shoot tip culture, the virus elimination effect of apical buds is generally _____ than that of the lateral buds, and the virus elimination effect of the buds in the vigorous growth period is _____ than that of dormant buds or buds that are about to enter dormancy.

7. During the removal of plant viruses, the combination of _____ and _____ can significantly improve the virus elimination effect, but falls short at _____.

## III. True or False Questions

( ) 1. The virus distribution in the plant is not uniform, and the apical meristem is generally virus-free, so virus-free plantlets can be obtained through an apical meristem culture.

( ) 2. The size of the shoot tip is positively correlated with the survival rate of shoot tip culture and the ability of differentiation and growth of stem leaves, but negatively correlated with the virus elimination effect.

( ) 3. Generally, the virus elimination effect of lateral buds is better than that of apical buds, and the virus elimination effect of buds in the vigorous growth period is better than that of dormant buds or buds that are about to enter dormancy.

( ) 4. Virus identification is not required for virus-free plantlets obtained by shoot tip culture or other means.

( ) 5. Studies showed that apical meristem culture can remove viruses from the phytoma and obtain virus-free plants.

( ) 6. Hot water treatment has a good effect on actively growing shoot tips, as it can not only eliminate the virus, but also provide a high chance of survival for the host plant. At present, this method is mostly used for heat treatment.

( ) 7. There is a possibility of reinfection of virus-free plantlets even if they are planted in the isolation area, so regular virus detection is still required.

## IV. Multiple Choice Questions

1. _____ of the following parts of the same plant carries the lowest viral content.

A. Stem segments  B. Root tip growing point cells

C. Stem node cells  D. Petiole cells

2. The most commonly used virus elimination method in plant tissue culture is _____.

A. Virus Elimination Method by Shoot Tip Culture

B. Virus Elimination Method by Callus Culture

C. Virus Elimination Method by *In Vitro* Micrografting Culture

D. Virus Elimination Method by Anther Culture

3. Because many viruses in cultured plants have a delayed recovery period, identification is still required at regular intervals during the initial _____ months.

A.18  B.12  C.10  D.24

4. In the test of the virus elimination effect of virus-free plantlets, _____ shows the best specificity and effectiveness.

A. Indicator Plant Method  B. Antiserum Identification Method

C. Electron Microscope Detection Method  D. Nucleic Acid Test

5. Long-term preservation of virus-free plantlets can be achieved by _____.

A. Low Temperature Preservation  B. Ultra-low Temperature Preservation

C. Isolation Preservation  D. Planting Preservation

## V. Short Answer Questions

1. Briefly describe the principle of virus-free plantlets obtained by shoot tip culture of plants. What are the factors that affect the virus elimination effect of plant shoot tip culture?

2. Analyze and compare the main features of various plant virus elimination methods.

3. Analyze and compare the main features of methods for the identification of virus-free plantlets.

# Module 5 Tissue Culture of Ornamental Flowers

## Source

The flower industry has become one of the pillar industries of agriculture with its rapid and sustainable development of China's economy and the gradual improvement of people's consumption level. Conventional methods such as seed, cutting, grafting, and division to propagate flower plantlets are far from meeting the needs of production. The tissue culture technique provides an economical and effective way for the rapid propagation of flower plantlets and the culture of virus-free plantlets, creating favorable conditions for the development of the flower industry.

This module mainly adopts the technique of tissue culture and rapid propagation of ornamental flowers such as *Phalaenopsis* orchid, carnation, tulip, and *Gerbera jamesonii*, so that students can master the technique of tissue culture and rapid propagation of flower plantlets, and consult for relevant medium formula to provide technical support for plantlet breeding of famous and rare flowers.

## Work Tasks

### Task 1 Tissue Culture of *Phalaenopsis* Orchid

*Phalaenopsis* orchid is a genus of *Phalaenopsis* in the family Orchidaceae and features a short stem (2–3 cm long) and large leaves. It has one or several pedicels, which are arch-shaped, sometimes branched, and waxy. It has large flowers with long flowering periods and bright colors(white and purplish-red, and yellow, light green or with purplish-red stripes on the petals). Known as the "Queen of Orchids", it is a rare tropical orchid. *Phalaenopsis* orchid is deeply admired by people for its high ornamental value as its beautiful and unique flowers resemble butterflies dancing. Mainly used for potted flowers or cut flowers, it occupies a large proportion of the orchid market.

*Phalaenopsis* orchid is monopodial epiphytic, with very few lateral branches growing on the plant, so it is more difficult to carry out conventional asexual propagation compared with other types of orchids. Its seeds do not contain endosperm and are difficult to germinate under natural conditions, so its own vegetative propagation and seed propagation are difficult to meet the needs of mass propagation. Tissue culture is an important means of rapid propagation.

## I. Selection and Sterilization of Explants

### (I) Culture of Pedicel Axillary Buds

1. Material Collection

*Phalaenopsis* orchid pedicels are used as explants when fresh from the emergence of pedicels to withered flowers. Take the whole pedicel, remove the petals, rinse it with tap water, then soak it in laundry detergent for 5–7 min, and rinse it repeatedly; or wipe the exterior of the pedicel with a cotton ball dipped in 75% alcohol.

**Knowledge Points**

The tissue culture of *Phalaenopsis* orchid usually adopts seeds, pedicel axillary buds, and pedicel dormant buds as explants. The seed regeneration pathway has been used for plantlet propagation of *Phalaenopsis* orchid and only for breeding purposes in recent years. However, the regeneration bud pathway originated from the explants of pedicel dormant buds or axillary buds promotes the dormant buds to transform into vegetative buds and directly proliferates by means of cluster buds, thus forming a callus without dedifferentiation and resulting in little variations of plants with relatively neat characters, and faster plantlet emergence compared with other ways. It is the main propagation pathway adopted in the propagation of commercial plantlets of *Phalaenopsis* orchid at present. The common technical routes of tissne culture of *Phalaenopsis* orchid are shown in Fig. 5-1.

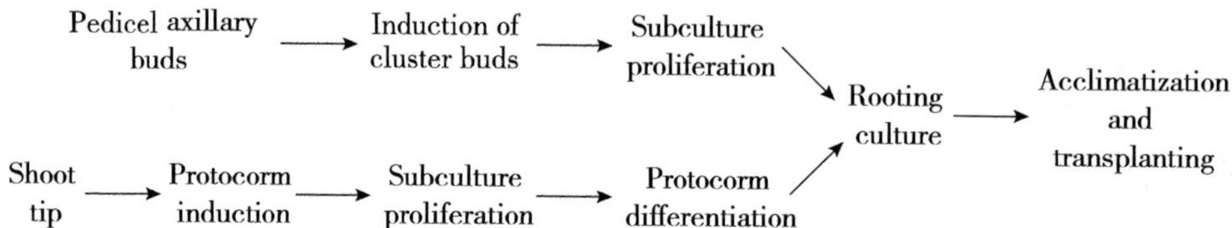


**Fig. 5-1 Common Technical Routes of Tissue Culture of *Phalaenopsis* orchid**

2. Disinfection and Inoculation

Cut the pedicel into about 2 cm segments by bud (1 bud for 1 segment) on the clean bench, remove the bract on the pedicel axillary bud, clean the joint between the segment and the bud with a scalpel, and cut off the diseased or wilted part on the surface of the pedicel with a scalpel. Soak the pedicel segments in 75% alcohol for 30 s. Then soak them in 0.1% mercuric chloride solution for disinfection for 10–15 min, repeatedly rinse them with sterile water 4–5 times, cut off 1–3 mm from both ends of the pedicel that had been in contact with sodium hypochlorite with a scalpel, inoculate them on the induction medium according to their growth direction, and inoculate 1 explant per flask, and place them in incubation room for culture.

**Knowledge Points**

The disinfection time varies with the texture and tenderness of the material. Generally, the disinfection time is 15–20 min for mature pedicels with blooming flowers or wilted flowers and 10–12 min for tender pedicels. The disinfection time should be appropriate. The disinfection is not complete when the application is too short and the explant will be damaged when the application lasts too long. The flask should be shaken frequently during disinfection to allow full contact of explants with the disinfectant and improve the disinfection effect.

3. Induction Medium and Culture Conditions

MS + 3.0–5.0 mg/L 6-BA + 2 g/L tryptone + 30 g/L coconut juice + 35 g/L sucrose + 12 g/L agar, pH 5.0–5.4. Culture at 25–28°C with 10–12 h illumination per day at a lighting intensity of 1,500–2,000 lx and air humidity of about 80%.

## (II) Shoot Tip Culture

1. Material Collection

*Phalaenopsis* orchid is monopodial epiphytic with rare lateral buds, so its lateral buds cannot be cut as explant like other orchids. Therefore, for the shoot tip culture of *Phalaenopsis* orchid, only the tip of plantlets or plants or the shoot tip of cluster buds induced by pedicel axillary buds can be cut as explants.

2. Sterilization and Inoculation

Take plantlets with 5–6 leaves, cut off leaves, rinse them with running water for 30 min, soak them in 75% alcohol for 30 s on a clean bench, and then soak them in 0.1% mercuric chloride solution for 7–10 min, rinse them repeatedly with sterile water 4–5 times, and cut off shoot tips (3 ± 0.1 mm) under sterile conditions and inoculate in the induction medium; or directly cut off shoot tips (3 ± 0.1 mm) of test-tube plantlets formed by pedicel axillary bud culture on a clean bench and inoculate in the induction medium to induce the protocorm.

**Knowledge Points**

The protocorm is a morphological structure in the germination of orchid seeds. In the initial stage of seed germination, no radicle is generated, but the proembryo germinates and expands, causing one end of the seed coat to rupture. The enlarged embryo takes the shape of a small cone, so it is called protocorm. Later, the tip of the protocorm is protuberant, resulting in the generation of scaly leaf primordia and the formation of cotyledon, which can be deemed as a shortened, bead-shaped organ composed of embryonic cells and resembling a tender stem.

Protocorm in plant tissue culture refers to a phenomenon unique to orchid tissue culture. It is an important form of rapid propagation technique for orchids. In the orchid tissue culture, protocorm specifically refers to those produced from the culture of shoot tips or lateral buds of plants such as orchids. More protocorms can be proliferated by cutting them into small pieces or causing injuries such as needling injury and transferring them to a fresh proliferation medium. After being cultured for a period of time, the protocorm gradually turns green and grows hairy rhizoids, and the leaf primordia develops into young leaves. Then it will be transferred and cultured for rooting to form complete regenerated plants.

3. Induction Medium and Culture Conditions

MS + 3.0–6.0 mg/L 6-BA + 15 g/L coconut juice + 20 g/L sucrose + 9 g/L agar, with a pH value of 5.4. The culture temperature is 25°C, the photoperiod is 10 h per day, and the lighting intensity is 1,500 lx.

## II. Subculture

### (I) Subculture for Cluster Buds

Usually, the axillary buds of pedicel are protuberant and enlarged after 7–10 days of culture and sprout after 15 days. Leaflets grow in about 30 days, and sprout, stretch and grow into individual plantlets with 4–5 leaves after 40–50 days. At this time, the plantlets grow normally, but the base of pedicel tissue, as well as the medium, turns black. The plantlet should be cut off the pedicel in time and transferred to MS + 3.0–5.0 mg/L 6-BA medium for subculture. After about 50 days, new cluster buds are generated with a multiplication coefficient of 3–4.

**Knowledge Points**

Browning refers to the phenomenon that the culture material releases brown substances into the medium, causing the medium and the culture material to gradually turn brown and perish. Browning occurs when polyphenol oxidases in the tissue are activated and phenolic compounds are oxidized to form brown quinones. Quinone compounds polymerize with proteins in the tissue of culture material under the action of tyrosinase, causing inactivation of other enzyme systems, metabolic disorders, and stunted growth.

The Main Factors Affecting Browning:

(1) Plant species and varieties. The amount of tannin and other phenolic compounds contained in plants varies, and the frequency and severity of browning also vary greatly. Generally, woody plants have a higher content of phenolic compounds than herbaceous plants and are more prone to browning, increasing the difficulty of tissue culture.

(2) Age, size and collection time of explants. The older the explants, the higher

the lignin content, and the easier they turn brown. The smaller the explants, the more susceptible they are to browning. Explant in the growing season contains more phenolic compounds and is prone to browning.

(3) Explant injury. The larger the cut of the explant, the larger the oxidized surface of phenolic substances, and the more severe the browning. Disinfectant and sterilizing agents can also cause browning by injuring the explants, e.g. sodium hypochlorite can cause browning in plants that are not susceptible to browning.

(4) Illumination. Shading treatment for the mother plant prior to material collection is effective in inhibiting browning, as many reactions in the oxidation process are controlled by enzyme systems.

### (II) Protocorm Subculture

After 2 weeks, the shoot tip expands and turns light green and hemispherical, and the diameter of the protocorm reaches 6 mm after 3 months. Cut the protocorm into several small pieces (2 mm × 2 mm) and transfer to a fresh medium; pieces being too small are easy to perish. The higher the inoculation density, the better the proliferation effect, showing the quorum sensing. The addition of 10%–15% coconut juice + 0.5–1 mg/L IAA + 5–10 mg/L 6-BA can accelerate the proliferation of protocorms.

**Knowledge Points**

The addition of 6-BA and NAA to the protocorm proliferation medium is more beneficial to the protocorm proliferation of *Phalaenopsis* orchid than using any of them alone. Low-concentration 6-BA helps the initiation of protocorm proliferation, while NAA helps the maintenance of the proliferation rate.

The addition of activated carbon to the medium can reduce protocorm contamination and calcification and prevent browning. The addition of potato juice can keep the protocorm's surface moist, which is conducive to proliferation and growth. The protocorm proliferation remains robust without the addition of hormones when colloid-containing fruit and vegetable juice is used to replace coconut milk.

## III. Rooting

Transfer protocorms not requiring subculture to rooting medium: 0.5 mg/L

MS + IAA or 0.3 mg/L MS + IBA, and add 20 g/L of sucrose, 40 g/L of coconut milk, 12 g/L of agar, and 5 g/L of activated carbon. After 60 days, the protocorms gradually differentiate, sprout, and develop into clustered plantlets. After 100 days, most of the plantlets grow 2–3 leaves. At this time, transfer the plantlets to the breeding medium again for rooting culture. After 70–90 days, they can grow into 4–5 cm high plantlets with 4–5 roots.

**Knowledge Points**

In the rooting stage, lower inorganic salt concentration facilitates the generation and growth of roots and the hormone ratio with lower cytokinin and auxin is appropriate. The combination of 0.1 mg/L of 6-BA and 0.5 mg/L of NAA is preferred, showing better root growth. An appropriate increase in lighting intensity can significantly promote plant growth. Under the optimum lighting intensity of 3,000 lx, the plant shows dark green in leaf color with thick leaves and robust plant growth. The addition of some natural extracts, such as potato juice, is beneficial to rooting. For example, the addition of 100 g/L of banana puree to the medium can significantly promote the root growth of cluster buds. The activated carbon with appropriate concentration also avails the rooting culture of *Phalaenopsis* orchid.

## IV. Acclimatization and Transplanting

Start acclimatization when plantlets grow to 4 cm high with 3–4 leaves and 3–4 roots. First, transfer test-tube plantlets into a greenhouse for 7–10 days. Second, uncap to acclimatize for 3–5 days. After taking out the plantlets and washing the culture medium at the root, rinse the culture medium and agar with clean water and do not injure the roots. Third, soak them into a low-concentration carbendazim or 0.05% potassium permanganate solution for 5 min, then take out the plantlets, dry up, and cool for 1h. Fourth, plant plantlets in the sphagnum plantlet tray and place in a cool and ventilated environment (the temperature is kept at 25–28°C; the initial humidity is about 85%, and then gradually drops to about 70%). After acclimatization, the lighting intensity is gradually increased to 6,000–8,000 lx in about 2 weeks, and the plantlet survival rate reaches more than 95%.

## Task 2 Tissue Culture of Carnation

*Dianthus caryophyllus*, also known as carnation and clove pink, is a perennial herb of *Dianthus*, Caryophyllaceae, Centrospermae, Dicotyledoneae, and Angiospermae. Originating from the Mediterranean region, carnation is mainly distributed in the temperate parts of Europe as well as in Fujian and Hubei, China, and is currently one of the best-selling fresh-cut flowers in the world. Its plant is 40–70 cm high, glabrous, and pinkish-green. Stems are cespitose and erect, with base lignified and upper parts sparsely branched. Leaves are linear and lanceolate with taper apex, short base, and distinct midvein, which is depressed above and slightly raised below; flowers are often solitary with fragrant branches in pink, purplish-red, or white, and the ovoid capsule is slightly shorter than persistent calyx.

Carnation should not be propagated by seed, as it can easily lead to mixed varieties, making it hard to maintain the good quality of the original varieties. In production, lateral bud cuttings are often used, but long-term vegetative propagation will cause serious virus infection, deteriorate the quality of cut flowers, and reduce flower yield, so the market demand cannot be met. Shoot tip tissue culture of carnation can not only keep the excellent parental traits and allow rapid propagation in large quantities in a short period of time, but also effectively remove viruses. It is an important measure in integrated virus control. The virus elimination method by shoot tip culture and tissue culture, as well as rapid propagation techniques of carnation have been used for the production of mother plants of carnation and are widely applied in the industrialized production of seedlings.

### I. Medium Formula

Induction medium: MS + 0.5 mg/L 6-BA + 0.2 mg/L NAA.

Proliferation medium: MS + 0.2 mg/L 6-BA + 0.2 mg/L KT + 0.1 mg/L NAA.

Rooting culture: 1/2 MS + 0.5 mg/L NAA.

### II. Collection and Sterilization of Explants

The apical buds of vegetative shoots are commonly used as explants in the plant tissue culture process of carnation. Select new shoots with robust bases, take them off, and strip leaves layer by layer until 1–2 pairs of young leaves remain; cut off stem segments and blade tips, keep apical buds at about 1.0 cm, and rinse them with tap water.

Put the above materials into a container, disinfect with 75% alcohol on the clean bench for 30 s, and rinse with sterile water 3–4 times; Sterilize with 2% sodium hypochlorite for 8–10 min, and rinse with sterile water 4–5 times.

**Knowledge Points**

Due to the serious diseases of carnation in the field, especially the widespread carnation virus disease. The purpose of tissue culture is mainly to eliminate virus disease by shoot tip culture and produce tissue culture plantlets as the mother plant for the cultivation of cuttings. Therefore, in the selection of materials, we should choose strong plants free of diseases and pests, with dark-green leaves, bright flowers, excellent commercial traits, and superior varieties as the plants for bud collection. Next, select new shoots with a thick, clean base, preferably those that sprout after the first removal of the apex.

## III. Inoculation

If the sampled plants are confirmed to be virus-free by virus detection, they can be directly inoculated on an induction medium after sterilization and disinfection. If the explants have not been detected for the virus or have been confirmed to carry the virus, the shoot tip explant should be stripped on the clean bench with the aid of a dissecting microscope, i.e. the outer tender leaves should be gently removed with a scalpel (the stem-leaf junction is quite tender, so care should be taken not to break the stem during the operation. Otherwise it will make the stripping difficult). Then, expose the shoot tip, cut off 0.3–0.4 mm growing point at the shoot tip with a scalpel, and quickly inoculate onto the induction medium. The operation requires considerable skill to prevent water loss and death of the shoot tip.

**Knowledge Points**

The virus elimination methods include virus elimination by heat treatment, virus-free shoot tip culture, callus culture, anther culture, chemical detoxification, and shoot tip micrografting.

The meristem is often virus-free, but some plants carry viruses in shoot tips. For instance, carnation mottle virus (CaMV) is found in 33% of materials in the 0.1 mm shoot tip culture of carnation. Heat treatment combined with shoot tip culture is adopted when neither heat treatment nor shoot tip culture is effective. Carnation streak virus, mosaic virus, and ringspot virus, among others, can be eliminated by treating carnation plantlets at 38–40℃ for 6–8 weeks, and then cutting 0.25–0.5 mm long shoot tips from them for shoot tip culture. Mottle virus can be eliminated by treating the plantlets at 38℃ for 30 days, and all viruses can be eliminated by treating the plantlets for 2 months. Although it has been reported that some viruses cannot be eliminated through heat treatment, this technique does work when used in combination with shoot tip culture.

## IV. Induction Culture

Maintain the temperature of the culture room at 25±2°C, the photoperiod at 12 h/d, and the lighting intensity at 2,000 lx.

Shoot tips will turn green 3–4 days after inoculation, enlarge one week after inoculation, and expand leaves 25–40 days after inoculation. The cluster buds are then subcultured in the induction medium. After the plantlets reach a certain number, some of them should be preserved at low temperatures, and others should be taken out for virus detection. Those without viruses will be further propagated.

**Knowledge Points**

Virus-free plantlets do not gain additional resistance to the disease and are likely to be reinfected soon. Therefore, once the virus-free plantlets are obtained, they should be well isolated and preserved. If these virus-free plantlets are kept well, they can be preserved and used for 5–10 years.

Two commonly used methods are preservation by isolation and *in vitro* preservation.

1. Preservation by Isolation

Viruses are transmitted by insect vectors, direct contact, and parasitic plants, among which transmission by insect vectors prevails. To prevent reinfection, virus-free plantlets are usually cultivated in an isolation room and covered with a 300-mesh net, with a mesh size of 0.4–0.5 mm. This can effectively prevent insects from entering and spreading viruses. In addition, the soil should be strictly disinfected, and pesticides should be sprayed in a timely manner to ensure that the plant materials are cultivated in conditions that are strictly isolated from viruses.

2. *In vitro* Preservation

It refers to the preservation of virus-free plantlets cultured *in vitro* under a low temperature of 1–9℃ and low lighting intensity. Under such conditions, the plant materials grow very slowly, and the medium only needs to be changed every 6 months or once a year.

## V. Proliferation Culture

Inoculate cluster buds about 1 cm long and proved to be virus-free on the proliferation medium for subculture every 30 days. The reproduction coefficient can be 4–6.

Original plantlets that have gone through virus-free induction should be cultivated as mother plants, and meanwhile, individual plantlets should be selected from them at an equal interval for flowering identification to ensure genetic stability before scions are cut to raise saplings.

**Knowledge Points**

Vitrification of test-tube plantlets is a phenomenon in which the plantlets are translucent and often abnormal in morphology. Vitrified plantlets are test-tube plantlets with abnormal leaf growth, transparent or translucent watery tender shoots, short swelling stalks, as well as chlorosis, shrinking, longitudinal curling, fragileness, and brittleness of leaves. Vitrified plantlets are the result of a kind of physiological disorder or physiological lesion during plant tissue culture. They are overhydrated and malformed, with slow growth and low differentiation and propagation coefficients. Their leaf epidermis lacks cuticle, wax coating, functional pores, and palisade tissue and only has spongy tissue. With a low dry matter content of cells, it is difficult to induce rooting and guarantee survival after transplantation.

Vitrification occurs easily in the carnation proliferation culture. The following preventive measures may be taken during the culture process.

(1) Select superior individual plants that are free from pests and diseases, can represent the traits of the variety and have outstanding characteristics for bud picking and virus detection.

(2) When selecting the hormone proportion of the culture medium, it is necessary to flexibly adjust it according to the characteristics of different varieties and by considering both the multiplication coefficient and the quality of plantlets. Research shows that the vitrification level increases significantly when the concentration of 6-BA or KT reaches 2 mg/L. The concentration of auxins used in the proliferation medium is generally 0.1–0.3 mg/L. If it is too high, calluses will be formed.

(3) The temperature required for tissue culture of carnation is lower than that of other flowers. Within 18–25℃, the growth of plantlets slows down with the decrease in temperature, but the quality of plantlets obviously improves and the vitrification of plantlets reduces. When the temperature is high, the plantlets show strong growth while thin and weak with an obviously high occurrence of variation, vitrification, or semi-vitrification. Under large diurnal temperature variations, water drops will form on the wall of flasks, contributing to the formation of vitrified plantlets.

(4) For varieties with strong growth and easy to vitrify, the mere use of fluorescent lamps or natural light tends to cause vitrification of plantlets. The vitrified plantlets can be reduced if they are cultured under fluorescent lamps for the first 10 days after inoculation and then transferred under a window and cultured with 4,000–8,000 lx scattered light for 2–4 hours every day. Fluorescent lamps may still be used when there is no sunlight to reduce vitrification. However, for varieties with poor growth, strong sunlight tends to inhibit growth and differentiation, causing leaf yellowing. Therefore, proper adjustment is needed in production according to the growth of varieties.

(5) Generally, when the humidity in the culture flasks is 70%–80%, the growth needs can be met. If the humidity is too high, it has a great impact on the growth of plantlets. If an impermeable sealing film is used and the humidity in the flasks reaches

100%, the vitrification of plantlets worsens for most carnation varieties. There is a risk of complete vitrification or even water-stained over successive generations, leading to complete failure of tissue culture. Therefore, in the tissue culture of carnation, only a permeable membrane can be used to seal the mouth of flasks.

## VI. Rooting

When the height of the test-tube plantlets reaches about 2 cm, normal small buds can be cut off and inoculated into the rooting medium. Calluses will be formed first at the basal incision, and radial roots will grow around the calluses after 10–15 days. The roots can grow to 1.5–2 cm long after 30 days, with a rooting rate of over 95%.

## VII. Acclimatization and Transplanting

Before transplanting, remove the cap and acclimatize the plantlets under scattered light in the culture room for 3 days. After that, take out the plantlets and wash them to remove the rooting medium; soak them in 0.1%–0.2% carbendazim solution for 2–3 min, plant the plantlets shallowly in the holes and press the substrate gently. A few plantlets with poor rooting or rootless plantlets can be transplanted after their base has been dipped in 1,000–2,000 mg/L of NAA.

The substrate is required to be loose with good air and water permeability, and perlite or vermiculite, river sand, and humus can be used by mixing them in a certain proportion. After being transplanted, the plantlets are covered with a membrane for moisturizing. After one week, the membrane is gradually removed for ventilation, and water is sprayed regularly every day for moisturizing. During the plantlet breeding period, nutrient solutions are sprayed 3–4 times a month to promote the widening of leaves, thickening of stems, and darkening of leaf color, thus improving the survival rate of plantlets.

# Task 3 Tissue Culture of Tulip

Tulip (*Tulipa gesneriana* Linn.) is a perennial herbaceous flower of the genus *Tulipa* in the family Liliaceae. The plants are erect with 3–5 lanceolate to ovate leaves. They are 20–40 cm high, about 15 cm long, and 2.5 cm wide. Flowers are bell-shaped, bowl-shaped, ovate, spherical, lily-shaped, and multi-petaled. Flowers are single- or multi-colored and can be in different shades of white, red, pink, purple, brown, yellow, orange, among others. Tulips are native to the Mediterranean coast, Central Asia, and Türkiye, and are very popular for their bright colors. Known as the "Queen of Flowers", they are widely cultivated as potted flowers, cut flowers, and garden flowers, and are quite popular around the world.

Tulips are traditionally propagated by bulb division, which shows a low propagation coefficient and slow propagation rate. In addition, during long-term cultivation, the quality of tulips is often affected by virus infection and other reasons, resulting in serious degradation of varieties and restricting tulip production. Plant tissue culture not only enables rapid propagation but also removes viruses.

## I. Culture conditions

The induction medium: MS + 2.0 mg/L 6-BA + 0.2–0.5 mg/L NAA.

Subculture medium: MS + 1.0 mg/L 6-BA + 0.1 mg/L NAA.

Rooting medium: 1/2 MS+0.1–1.0 mg/L NAA +2% sucrose.

Adding 0.2% activated carbon to the induction medium can prevent browning. Adding 3% sucrose and 0.7% agar to each type of medium can achieve a pH level of 5.8. The culture temperature is 24–26°C, the lighting intensity is 2,000 lx, and the photoperiod is 12 h per day.

## II. Material Collection and Sterilization

Remove the outer skin and 1–2 layers of external scales of the bulb, cut a small amount of top and base tissue, rinse them with liquid detergent or washing powder, and then under running tap water for 30 min. Under aseptic conditions, soak them in 75% alcohol for 30 s and rinse with sterile water 3–5 times. Then, disinfect them in 0.1% mercuric chloride solution for 8–10 min and rinse with sterile water 3–5 times.

**Knowledge Points**

The scales, bulbs, flower stems, or anthers of tulip can all be used as explants. Among them, scales show the highest frequency of callus induction, followed by leaves and bulbs. The callus formation rate of the middle layers of scales is higher than that of the outer and inner layers of scales. The highest frequency of callus induction is found at the base of the scales near the bulblet, which can be used as a preferable material for callus induction in primary culture.

The selected materials can be first treated under a low temperature of 4℃ for about 30 days, which is beneficial for improving the survival rate and frequency of callus induction of plantlets through explant culture.

## III. Primary Culture

Before the sterilized scales are cut into small pieces of about 5 mm in size, their edges are removed. Then, the cut materials are inoculated into the induction medium with the convex surface upwards. Callus formation can be induced in about 20–40 days.

**Knowledge Points**

Plant regeneration generally comes into five types: sterile spur regeneration, organogenesis, cluster bud regeneration, embryogenesis, and protocorm regeneration. The plants formed in such ways are called regenerated plants.

Regeneration of tulip plantlets by inducing callus with bulb scales is called organogenesis. That is, the explants are induced to dedifferentiate to form calluses, and then the callus cells differentiate to form adventitious buds (cluster buds), which is also called callus regeneration. Callus regeneration can be further divided into the following four types: The calluses differentiate into only buds or roots, i.e. roots without buds or buds without roots. In most cases, buds are formed first, and then roots grow at the base of the stem after the buds elongate to form small plantlets. Roots are formed first and then buds emerge from their base to form plantlets. which are rare in monocotyledons and more common in dicotyledons. Buds and roots are respectively formed in different adjacent parts of the calluses, then they are combined to form plantlets. This is similar to the natural grafting of roots and buds but is rarely seen. Only when the vascular bundles of the buds and roots are connected can viable plantlets be formed.

## IV. Subculture

Place the induced callus on the proliferation medium, and it will differentiate into adventitious buds after culture for 30–40 days. Continue the proliferation culture, and the adventitious buds will further differentiate into new ones, but this takes a long time. Bulblets will gradually grow at the base of the buds with the extension of culture time.

**Knowledge Points**

In tissue culture, callus refers to a mass of undifferentiated tissue formed inside an explant or on the surface of its incision that has the ability to redifferentiate. Under certain culture conditions, the callus can form a bipolar embryoid through embryogenesis, or form a unipolar bud or root through organogenesis, and then form a complete plant again. This latter process is generally called redifferentiation.

A typical callus formed from a single cell or explant generally goes through three phases: initiation, division, and differentiation.

The initiation phase, also known as the induction phase, refers to the period when cells are ready to divide. Although the size of the cells does not change much during this period, physiological and biochemical changes occur inside the cells, such as enhanced anabolism and synthesis of proteins and nucleic acids.

The division phase refers to the process of proliferation, with one cell dividing into two daughter cells. Once the cells of the explants are induced, their outer cells begin to divide, allowing the cells to dedifferentiate. Calluses in the division phase are characterized by fast cell division, loose structure, lack of tissue structure, and light and transparent color.

The differentiation phase is when cells that have stopped dividing undergo physiological and metabolic changes to form a callus composed of cells with different morphology and functions. If the plant callus, once formed, is transferred into a differential medium, it will enter the differentiation and disintegration phase. In the telophase of cell division, a series of morphological and physiological changes begin to occur inside the cells, resulting in morphological and physiological differentiation of the cells and the emergence of cells with different morphology and functions.

## V. Rooting Culture

When the buds grow to 1.5–2.0 cm high, cut them off and transfer them into a rooting medium for rooting culture.

Generally, root differentiation begins 10 days later when the buds can be taken out of the flasks for transplantation. Tulip plantlets should be transplanted as soon as they have a few roots. The roots will turn brown if they are cultivated for too long, which is not conducive to the growth of the plantlets.

## VI. Acclimatization and Transplanting

When the test tube plantlets are 2–3 cm high with 3 or 4 new roots, they can be transplanted. Before transplanting, open the cap and acclimatize the plantlets for 2–3 days. Then, take them out and wash them to remove the root medium. After that, transplant them into the seedbed with perlite as the cultivation substrate, and shade them with a sunshade net in the early stage to maintain 20% temperature and 80% relative air humidity. With these measures, the plantlets can survive after 10 days.

## Task 4 Tissue Culture of *Gerbera jamesonii*

*Gerbera jamesonii*, also known as Wedding Flower, is a perennial evergreen herb that belongs to the genus *Gerbera* of the family Compositae. Native to South Africa, *Gerbera jamesonii* was introduced to Britain in 1878 after it was discovered by the British in the Transvaal region of South Africa, and gradually spread to the whole world. *Gerbera jamesonii* has large flowers in varied bright colors and straight flowering branches, promising a high percentage of cut flowers. It tends to flower continuously and annually in warm areas. As a result, it has developed rapidly in the international cut flower market and has become one of the world's four most famous cut flowers.

The natural setting rate of *Gerbera jamesonii* is low, and the seeds are easy to fail in germination (the germination rate is generally only about 15%). Seed propagation is not suitable for cut flower production on a large scale. In contrast, division propagation is slow, and tends to cause degradation and virus infection of plantlets, which cannot meet the market demand either. Therefore, tissue culture is generally adopted at home and abroad to produce *Gerbera jamesonii* plantlets. This technique features rapid propagation of high-quality plantlets on a yearly basis.

## I. Material Collection, Sterilization, and Inoculation of Explants

Since the shoot tip of *Gerbera jamesonii* is quite difficult to strip and easily contaminated, the receptacle is often used as an explant. Cultivars with large, colorful flowers that are popular on the market should be selected, and the culture material should be selected from superior individual plants that grow vigorously with flowers in authentic color and are free of diseases and insect pests. The preferred culture material should be flower buds with a diameter of 0.5–1 cm on stalks with a length of 1–3 cm. Flower buds being too big or small cannot obtain satisfactory results.

First, rinse the flower buds with tap water. Second, soak them in 75% alcohol for 30 s on the clean bench, and rinse them with sterile water 3–4 times. Third, disinfect them with 0.1% mercuric chloride solution for 15–20 min, add several drops of Tween. Fourth, rinse them with sterile water 3–4 times, and drain the water for later use.

On the clean bench, remove the bracts and all florets of the flower buds with forceps or a scalpel, leaving only the receptacle. Cut the receptacle into sections with a length of 0.2–0.3 cm and inoculate them on induction medium (MS+ 2.0 mg/L 6-BA + 0.2 mg/L NAA).

**Knowledge Points**

Plant organs and tissue materials collected from nature and greenhouses carry a variety of microorganisms. Once these contaminants enter the culture flasks, they will contaminate the culture medium and materials. Various sterilizing agents can be used to sterilize explants. Since different plants or different organs and tissues of the same plant vary in the number of microbes carried and in difficulty in sterilization, the type and concentration of sterilizing agents and the sterilization method also vary.

1. Sterilization of Shoot Tips, Shoot Segments, and Leaves

Before sterilization, rinse the explants with water or liquid soap, and then water for those with many fuzzes. Dry the surface of explants with absorbent paper, and soak the explants in 0.1%–0.2% mercuric chloride solution for 2–10 min, or in 70% alcohol for a few seconds and then in 10% calcium hypochlorite solution for 10–20 min, or in 70% alcohol for a few seconds and then in 2% sodium hypochlorite solution for 15–30 min for sterilization. After sterilization, pour out the sterilization solution, rinse the explants with sterile water 3–5 times, place the explants on sterile filter paper to dry the surface, and inoculate them after they are properly cut.

2. Sterilization of roots, tubers and bulblets

These materials grow in the soil where there are a great number of diverse microbes, and they are more likely to be damaged during digging, making them quite difficult to sterilize. Therefore, care should be taken to clean them carefully before sterilizing. The dents and the gaps between scales should be cleaned with a soft brush, and the damaged parts should be removed. During sterilization, time should be extended or the concentration of the sterilizing agent should be increased. For example, the explants should be soaked in 0.1%–0.2% mercury chloride solution for 5–12 min, or in 70% alcohol for several seconds, and then soaked in 6%–10% sodium hypochlorite solution for 5–20 min for sterilization. The operation steps after sterilization are the same as above.

3. Sterilization of Flower Buds

In the unopened flower buds, the anthers are wrapped by the floral envelopes and are in a sterile state, so they can be sterilized directly after being picked. During sterilization, soak them in 70% alcohol for 10–30 s, and then in 0.1% mercury chloride solution for 3–10 min or in 1% sodium hypochlorite solution for 10–20 min. After sterilization, rinse them with sterile water 3–5 times and take out the internal anthers for inoculation.

4. Sterilization of Fruits and Seeds

Some of these materials have fuzz or wax on them, and several drops of Tween-80 should be added to the sterilizing agent to increase the sterilization effects. After explants are rinsed with clean water and their surfaces are dried, Soak the fruits in 70% alcohol for several seconds, and then in 2% sodium hypochlorite solution for 10–20 min, or in saturated bleaching powder supernatant for 10–30 min for sterilization. Soak seeds in 10% sodium hypochlorite solution for 20–30 min, or in 0.1%–0.2% mercury chloride solution for 5–10 min for sterilization. After that, rinse the seeds 3–5 times with sterile water, and take out the tissue of the fruit or the seed for inoculation.

## II. Primary Culture

For the culture condition, the temperature is 25+2°C, the lighting intensity is 2,000–3,000 lx, and the photoperiod is 12–14 h. Culture in darkness for 2–3 days after inoculation before culture under normal conditions can effectively reduce the browning of explants. Yellowish-white calluses are gradually formed at the wound 7–10 days after receptacle inoculation, and most of them gradually turn green after 15 days.

## III. Adventitious Bud Induction Culture

Cut the green callus into small pieces and inoculate on the adventitious bud induction medium (MS+1.0–5.0 mg/L 6-BA+0.2–0.5 mg/L NAA).

For some cultivars, adventitious buds may be induced after one month. For most cultivars, adventitious buds appear until 3–5 months after continuous subcultures; For a few cultivars, no adventitious buds are induced at all.

## IV. Subculture

Since the probability of inducing adventitious buds from the receptacle is relatively small, any adventitious buds, once formed, should be promptly transferred to subculture medium (0.2–1.0 mg/L MS+6-BA + 0.05–0.1 mg/L NAA for rapid expanded propagation).

For initial subculture propagation, the concentration of 6-BA can be increased to 2.0–3.0 mg/L, considering the small base number of adventitious buds. As the base number continues to increase, the concentration of 6-BA should be gradually reduced to prevent the production of vitrified plantlets.

**Knowledge Points**

Cytokinin was discovered by Skoog et al. (1948). It is synthesized at the root tip and transported upwards, and its main effects are to promote cell division and organ differentiation, promote lateral bud differentiation and growth, inhibit apical dominance, and retard tissue senility. In plant tissue culture, in addition to KT and ZT, 6-BA, Thidiazuron (TDZ), and 2-isopentennyladenine (2-ip) are commonly used cytokinins with the same effects. Among them, TDZ is more effective in inducing adventitious buds, but it also easily causes vitrification of the culture. Cytokinins are used in the concentration range of $10^{-7}$ to $10^{-5}$ mol/L. Cytokinins are added to the medium with the aim of promoting cell division and inducing calluses, adventitious buds, and adventitious embryos. A high concentration of cytokinin inhibits the production of roots. Among the cytokinins, ZT is more expensive and easily destroyed in autoclaved sterilization, while 6-BA and KT have stable properties and are relatively cheap. However, ZT is more effective in the induction of adventitious embryos in some plants.

## V. Rooting Culture

When the number of adventitious buds of *Gerbera jamesonii* reaches a certain base after expanded propagation and subculture, Cut off the individual plantlets 2–3 cm high and transfer them into a rooting medium (1/2 MS +0.1 mg/L NAA or 0.3 mg/L IBA) for rooting culture. After 7–8 days, 3–5 adventitious roots will grow at the base of the plantlets, and the rooting rate can be as high as 98%. After 12–15 days, when the root length reaches 0.8–1.5 cm, the plantlets may be taken out of the flasks for acclimatization. If the roots become overly long, it is not conducive to acclimatization and transplanting.

Auxins with a low concentration, preferably within 0.1 mg/L of NAA are used in the rooting phase. If the concentration of auxins is high, such as above 0.5 mg/L of NAA, the root system will be short, coarse-textured, and form calluses, making it easy to fall off and rot during transplantation.

**Knowledge Points**

As early as 1926, Went, a Dutch scholar, discovered a substance that could promote the growth of oat coleoptile. In 1930, it was confirmed that this substance was IAA. Auxins are synthesized at the shoot tip and transported down the plant body to promote cell growth and rooting. After the discovery of IAA, substances with similar effects, such as NAA, 2,4-D, and IBA were successively synthesized and widely used. In plant tissue culture, auxins play an important role in inducing calluses and roots. 2,4-D also induces the formation of adventitious embryos in the tissue culture of some plants. The combination use of auxins and cytokinins can play a regulatory role in organ formation and plant regeneration. However, since IAA can be destroyed during autoclaved sterilization, sterilization by filtration is preferred.

### VI. Acclimatization and Transplanting

For transplanting, transfer the plantlets in the flasks under scattered light for 3–5 days of acclimatization, and then take out and wash to remove the root culture medium before transferring them to the seedbed. The medium may be perlite, vermiculite, and coarse sand, among others. Care should be taken to ensure shading and moisturizing. The plantlets will begin to form a new root system 15 days after being transplanted. At this time, lighting can be increased gradually and fertilization can be performed. When the plantlets grow well with 2–3 fresh leaves, they may be transplanted again.

## Task 5 Tissue Culture and Rapid Propagation of *Haworthia* Duval

*Haworthia* Duval represents more than 150 species of small perennial plants of Liliaceae, mostly in South Africa and Southwest Africa. Named in memory of Adrian Hardn Haworth, a British botanist, it also has another name called *Lepisorus*. However, this name is confusing as it sounds like that in the Polypodiaceae of ferns. Therefore, it is mostly called *Haworthia* Duval. The leaves of *Haworthia* Duval are lance-shaped triangular with full and glossy mesophyll. There are colorful stripes or star-shaped spots on the light green leaves. The leaves can be hard or soft. The surface of hard leaves often has white spots, or stripped and fumulus-like nodules, such as *Haworthia fasciata*. Soft leaves have a more distinct "window-shaped" structure with more plump and transparent mesophyll and vellus on the edges, such as *Haworthia cooperi* var. *truncata*. With a unique, cute appearance and small size, they are popular and well-received on the market and are often used for indoor decoration. They are not only delightful but also can purify the environment, contributing to environmental protection.

Because of a small population of *Haworthia* Duval plants in Asia and a low reproductive rate, our horticultural researchers have been introducing excellent species of *Haworthia* from overseas and have conducted massive research in achieving rapid propagation of the plants. Initially, vegetative propagation methods such as traditional leaf-cutting and base propagation were used, but these methods show such a low propagation coefficient and low efficiency that the market demand cannot be met. In addition, they tend to damage the mother materials. In recent years, horticultural researchers have gradually started to use the tissue culture technique to achieve rapid propagation of superior cultivars, and have obtained remarkable effects. This technique can not only realize the rapid propagation and culture of plantlets but also effectively retain the ornamental features of superior cultivars. In recent years, progress has been made in the research on the techniques for tissue culture and rapid propagation of *Haworthia* Duval, including the selection of explants, minimal media, hormones, and other additives, measures for prevention and control of explant browning, and techniques for rooting and culture of strong plantlets.

## I. Selection of Explants

In the tissue culture of *Haworthia* Duval, inflorescences and leaves are usually used as explants as they show a relatively high rate of success in induction. Although the leaves have a weak ability to divide and a long differentiation period, the frequency of callus induction and the differentiation rate are high. A rapid propagation system may be established. Inflorescences are the most suitable explants since they are convenient to collect, cause no harm to the mother materials, and can be effectively sterilized with a high frequency of callus induction and a high cluster bud differentiation rate. In addition, the plantlets obtained in this way completely retain the excellent characteristics of the mother materials.

In recent years, there has been more and more literature about succulent research at home and abroad, in which researchers have experimented with a variety of explants for tissue culture, such as seeds, new lateral buds, inflorescences, leaves, and ovaries. It has been shown that the frequency of callus induction varies considerably in the tissue culture of *Haworthia* Duval plants with different explants. Initially, Ogihara from abroad used flower buds as explants to successfully induce calluses, and later, foreign researchers also realized successful induction using intact leaves as explants. Subsequently, Xu Jiyong et al. in China used scape as an explant in their study of tissue culture and rapid propagation of *Haworthia truncata* Schönland, and achieved successful induction without damaging the mother materials. Wang Quan et al. successfully induced calluses by using the spring-grown ovary part as the explant in the experiment of tissue culture and rapid propagation of *Haworthia emelyae* var. *major*, but vitrification easily occurred in the process. In general, flower stems and ovaries are used in the tissue culture and rapid propagation experiments of most plants. However, there are limitations in using young pedicels or ovaries as explants in the tissue culture and rapid propagation system of *Haworthia* Duval plants, as the mother plant is seasonal and needs to grow to a certain age before flowering. Meanwhile, it has been reported that new lateral buds, seeds, scapes, inflorescences, and leaves can also be used as explants, but they all have some limitations to a greater or lesser extent. For example, new buds growing in the roots may be poorly sterilized and easily contaminated; seeds are hybrid products, and the plantlets obtained are more variable compared with the mother materials.

## II. Selection of Culture Media

In the tissue culture, culture media provide the necessary basic nutrients for plant growth and differentiation. The composition of culture media required by different explants in different growth stages is also different. For example, the requirements for culture media are not the same in the stages of inducing and differentiating calluses and adventitious buds, subculture, and propagation culture. The optimal minimal media for *Haworthia* Duval plants are MS and 1/2 MS.

In the tissue culture and rapid propagation experiment of *Haworthia cooperi* and *Haworthia cooperi* var. *obtusa*, the inflorescence was used as the explant, and it was concluded that the optimal minimal medium for the induction and differentiation of calluses in *in vitro* rapid propagation of *Haworthia cooperi* was MS, while the optimal minimal medium for the rooting stage was 1/2 MS.

## III. Application of Hormones

In general, the hormones commonly used in tissue culture of the *Haworthia* Duval plants are KT, 6-BA, NAA, IAA, IBA, and AC.

Propagation and differentiation of explants of the *Haworthia* Duval plants can be achieved on a minimal medium, and adding hormones and other additives will promote this process. Since plant growth regulators can directly or indirectly regulate plant biochemical processes to a certain extent, and even realize the precise control of growth in each stage, care must be taken in the use of plant growth regulators.

Hormones of different types and with different concentrations have different effects on induction, differentiation, propagation, growth, and rooting. According to tissue culture studies of multiple *Haworthia cooperi* carried out by Guo Shenghu et al., when 6-BA was selected as the regulatory hormone, a low concentration of auxin helped the differentiation of cluster buds in the process of callus induction, while a high concentration of growth hormone led to the vitrification of cluster buds. For plantlet growth and rooting, IBA was better than NAA for *Haworthia cooperi*, and if both IBA and NAA were added with a concentration of 0.5 mg/L and 0.1 mg/L respectively, the rooting effect was the best. Compared with common *Haworthia cooperi* var. *venusta*, they have a larger and more transparent "window-shaped" structure. The research by Zhang Jingxin et al. confirmed that different hormone combinations had different effects on the elongation of adventitious buds of *Haworthia cooperi* (Ice Lantern). The growth rate of plant stems will increase with the improvement of IAA concentration. When IAA concentration reaches 0.5 mg/L, the height of the stem reaches the maximum. As the hormone concentration continues to increase, the growth rate of the stem gradually decreases. At this time, the leaf thickness gradually increases, and vitrification occurs. Therefore, when the ratio of IAA to BA is too large, it is not conducive to the rapid and robust growth of the stem. Additionally, the growth rate and differentiation of stems are not as good as those treated with low-concentration BA when a fixed amount of high-concentration BA is used in combination with IAA with different concentrations.

In the research on the effects of different hormones on the morphological characteristics of *Haworthia cooperi* (Ice Lantern), Zhang Hailong et al. found that the plantlets obtained through tissue culture basically failed to grow normally if no MS medium was added. In addition, adding a certain amount of growth hormone IAA still failed to improve the situation, but instead led to fleshy roots and even vitrification. However, when a certain amount of BA was added, the situation changed and the plantlets started to grow normally. The conclusion of this experiment is consistent with those of the above experiments of Guo Shenghu et al. and Zhang Jingxin et al. and confirms the value and importance of hormones in tissue culture.

## IV. Culture Conditions

The external culture conditions are also an important factor affecting the *Haworthia* Duval plants. Lighting intensity has a significant effect on the proliferation of cultured cells and the differentiation of organs. Generally speaking, the higher the lighting intensity is, the more robust the plantlets grow, while on the contrary, the plantlets will grow excessively when the lighting intensity is low.

The research by Zhang et al. shows that the growth rate and morphology of plantlets in tissue culture vary greatly under different levels of lighting intensity. When the lighting intensity reaches 3,000 lx, the excessive growth of plantlets is eased, i.e. they grow at a reduced rate. In terms of morphological recovery rate, the effect is the best when the lighting intensity is 3,000 lx and the temperature is 20°C. In addition, plantlets obtained through tissue culture also grow at a fast rate.

## V. Preventive Measures for Explant Browning and Vitrification

*Haworthia* Duval plants are prone to browning at the incision of the explants. Proper shading is effective in suppressing the browning; while minimizing the area of explant incisions and keeping them flat can also reduce browning to a certain extent. During induction and differentiation in tissue culture, vitrification of calluses and cluster buds often occurs. For *Haworthia* Duval plants, vitrification is serious when the concentration of 6-BA is high. For cluster buds, this problem can be basically solved when the concentration of 6-BA ≤ 0.5 mg/L and NAA is adjusted to 0.01 mg/L.

Plant tissues contain PPO, which reacts with phenolic substances to form quinone, resulting in browning at the incision of explants. Therefore, whether explant browning can be effectively controlled is critical for achieving success in tissue culture and rapid propagation. According to the research by Gu Yanze et al., proper shading is effective in suppressing the explant browning. Moreover, according to the findings of Qin Liyuan et al., minimizing the area of explant incisions and keeping them flat can also reduce browning to a certain extent. During induction and differentiation, the second critical step in tissue culture, calluses and cluster buds often show vitrification, which leads to the failure of tissue culture and rapid propagation. Vitrification, also known as overhydration, is related to the explants selected, the pH value of the minimal medium, the type and concentration of hormones, and the light and temperature conditions.

## VI. Rooting and Growth Promotion

When the cluster buds in the medium grow into normal plantlets, the growth promotion stage begins. The medium formula at this stage will directly determine the rooting level of plantlets. Currently, the common rooting medium is 1/2 MS minimal medium with hormones such as NAA, IBA, and 6-BA added.

The rooting and growth promotion stage is the key to the survival of transplanted plantlets. Therefore, the type and concentration of hormones in the rooting medium for growth promotion have become the focus of research in the industry. Currently, commonly used hormones mainly include NAA, IBA, and 6-BA. According to the research by Wang Ting et al., adding 2 mg/L of IBA to 1/2 MS as the medium is most beneficial to the rooting of callus-induced buds of *Haworthia cooperi* var. *truncata*. In addition, the best medium for the rooting experiment of *Haworthia maughanii* is 1/2 MS + 0.5 mg/L NAA, which is confirmed through the experiments by *Li Yuan* et al. Similarly, according to the adventitious root formation experiment of *Haworthia cooperi* (Ice Lantern) performed by Zhang Jingxin et al., the best rooting medium is MS + 0.2 mg/L NAA + 20 g/L sucrose. After 30 days of culture on the medium, the rooting rate will reach 100%, with the highest number of rooted individual plants and a strong root system. However, in the rooting and growth promotion experiment of *Haworthia cooperi* performed by Guo Shenghu et al., it was concluded that the rooting effect of IBA on *Haworthia cooperi* was better than that of NAA, and that the best rooting effect can be achieved with 0.5 mg/L of IBA and 0.1 mg/L of NAA. This result supported the research conclusions of Wang Ting et al., and also verified the research results of Li Yuan et al. and Zhang Jingxin et al.

## VII. Existing Problems and Development Prospects

In recent years, although more and more studies have been conducted on tissue culture and rapid propagation of *Haworthia* Duval at home and abroad, with certain achievements made, bottleneck problems of rapid propagation have not yet been solved for many varieties of *Haworthia* Duval, and problems such as low reproduction rate, vitrification, and browning are yet to be tackled in a better way. Moreover, there are so many influencing factors, such as varieties, explants, growth regulatory hormones, other additives, and external culture conditions, resulting in many differences in research direction and results. To some extent, this reflects that different varieties and explants have varying requirements in terms of the culture media and culture conditions. Therefore, more detailed research is required in this field so as to establish a better scientific, theoretical system to better achieve mass and high-quality reproduction of rare species of *Haworthia* Duval, so that the native species and rich variety resources of *Haworthia* Duval can be well retained and protected. In addition, in recent years, in spite of many studies conducted on the tissue culture, rapid propagation, and cultivation physiology of *Haworthia* Duval, only a few of them focused on the selection and breeding of new varieties. The selection and breeding of varieties are still dependent on the traditional hybridization mode. As a result, most of the new varieties of *Haworthia* Duval in China are introduced from abroad, resulting in a limited choice of varieties. Therefore, it is necessary to strengthen the breeding of *Haworthia* Duval, especially by using new breeding methods such as mutation breeding and molecular marker selection to cultivate new varieties with good ornamental value, fast growth, and easy cultivation.

Compared with the traditional way of division propagation, tissue culture and rapid propagation are not restricted by natural conditions, which increases the multiplication coefficient and propagation speed of succulent plants in the shortest time, regardless of the influence of season and climate. The plantlets of superior varieties can be reproduced efficiently with less energy consumed all year round. This technique also helps to solve the problem of extinction of rare varieties of *Haworthia* Duval and makes it possible to meet the market demand. Tissue culture realizes industrialized production of succulent plantlets with good economic benefits and market prospects, which can meet the current requirements of rapid and efficient agricultural modernization for a large number of high-quality plantlets, and guide the future reproduction of succulent plants in a new direction. It also has important practical significance in the preservation of superior germplasm resources of succulent plants, propagation of famous, superior, and rare varieties, and rapid propagation for export.

# Summary

Browning, contamination, and vitrification are known as the three major problems in plant tissue culture.

Browning refers to the phenomenon that the culture materials release brown substances into the medium, causing the media and the culture materials to turn brown and perish gradually. Browning occurs when the polyphenol oxidases in the tissue are activated and the phenolic compounds are oxidized to form brown quinones. Quinone compounds polymerize with proteins in the tissue of culture materials under the action of tyrosinase, causing inactivation of other enzyme systems, metabolic disorders, and stunted growth.

The main factors affecting the occurrence of browning are plant species and varieties; age, size, and sampling time of explant; explant damage; lighting.

Vitrification of test-tube plantlets is a phenomenon in which the plantlets are translucent and often abnormal in morphology. Vitrified plantlets are test-tube plantlets with abnormal leaf growth, transparent or translucent watery tender shoots, short swelling stalks, and chlorosis, shrinking, longitudinal curling, fragileness, and brittleness of leaves.

The prevention of vitrification of carnation plantlets should focus on the following aspects: selection of explants, hormone ratio of culture medium, temperature, lighting, and humidity.

The two common preservation methods of virus-free plantlets are isolation preservation and *in vitro* preservation.

Auxin plays an important role in the induction of callus formation and rooting, and also induces the formation of adventitious embryos. The use of auxins in combination with cytokinins can play a regulatory role in organ formation and plant regeneration.

Cytokinins are added to the medium with the aim of promoting cell division and inducing calluses, adventitious buds, and adventitious embryos. A high concentration of cytokinin inhibits the production of roots.

# Review Test

## I. Fill in the Blanks

1. The explants of succulent plants for tissue culture and rapid propagation generally include_______, _______, _______, and_______.

2. For tissue culture of *Haworthia* Duval, _______and_______are usually used as explants as they show a relatively high rate of success in induction.

3. The optimal minimal media for *Haworthia* Duval plants are_________and _______.

4. _________and_________can effectively reduce the browning in the tissue culture of *Haworthia* Duval plants.

## II. Short Answer Questions

1. Describe the current status and problems of tissue culture of succulent plants.

2. Briefly describe the tissue culture process of *Haworthia cooperi.*

3. Briefly describe the selection of explants for tissue culture of *Phalaenopsis* orchid.

4. What is the common technical route for plant tissue culture of *Phalaenopsis* orchid?

5. What is browning? What are the factors that affect its occurrence?

6. What is vitrification? What measures are taken to prevent the vitrification of carnation?

7. What are the common virus elimination methods for carnation?

8. What are the common methods and precautions for plant tissue culture of tulip?

9. What are the common plantlet regeneration methods?

10. What are the roles of auxins and cytokinins in plant tissue culture?

11. What are the sterilization methods for different explants?

# Module 6 Tissue Culture of Fruit Trees

## Source

Berry fruits, such as blueberries and grapes, all have high nutritional value, and some of them also have high medicinal value. For example, raspberry and *Actinidia arguta* are popular fruits of the third generation at home and abroad in recent years. Superior fruit tree varieties cultivated through cutting propagation cannot meet the market demand. In contrast, the plant tissue culture technique is characterized by fast propagation of virus-free plantlets with no changes in the variety trait and is not restricted by planting locations and seasons. This technique is now widely used in the rapid propagation of fruit tree plantlets.

## Work Tasks

### Task 1 Blueberry Tissue Culture

Blueberries (*Vaccinium ashei*) classified in the family Ericaceae within the genus *Vaccinium*, is a perennial deciduous or evergreen shrub. Blueberries have good health benefits and high medicinal value. They contain anthocyanin, vitamins A and C, ellagic acid, and other antioxidant substances, which have the effects of relieving eye fatigue, improving vision, delaying brain nerve aging, enhancing anti-cancer immunity and memory, boosting the immune system, and preventing cardiovascular diseases. Ranking among the top five health foods, they are called the king of fruits in the world. In addition, because of their high economic and commercial value, blueberries have drawn wide attention from consumers at home and abroad.

With the growing demand for blueberry plantlets, blueberries show broad market prospects. At present, blueberries are mainly bred by cutting propagation, seed propagation, grafting, and tissue culture and rapid propagation. However, as seed propagation, grafting, and cutting propagation have shortcomings such as low germination, and survival rate, small multiplication coefficient, and long cycle, the market demand still cannot be met. With the tissue culture and rapid propagation technique, propagation becomes faster and the traits of varieties can be maintained. This is currently one of the most important techniques for the mass propagation of blueberries. In this task, blueberry shoot segments are used as the main experimental materials for tissue culture and rapid propagation of blueberry plantlcts.

## I. Medium Preparation

Primary culture medium: Modified WPM medium + 2% sucrose + 0.5% agar powder.

Subculture medium: Modified WPM + 2.0 mg/L ZT+ 2% sucrose + 0.5% agar powder.

Rooting medium: 1/2 modified WPM+1.0 mg/L IBA+ 2% sucrose+ 0.5% agar powder+ 0.1% activated carbon.

## II. Selection and Sterilization of Explants

Cut the tender shoots from healthy, disease-free, and pest-free blueberry plants of the current year; rinse under running water for 2 h, and cut into 3–4 cm long segments with buds. On the sterile clean bench, sterlize the segments first with 75% alcohol for 30 s, and rinse with sterile water 3 times. Then, sterilize with 0.1% mercuric chloride for 6–8 min, and rinse with sterile water 4–5 times. Finally, dry up with sterilized filter paper for inoculation.

### Knowledge Points

The selection of explants is of great importance to tissue culture and is a primary concern in establishing a rapid propagation system. At present, many types of explants have been selected for blueberry tissue culture, including leaves, tender shoots, shoot segments, dormant branches, green branches, seed embryos, and anthers, among which shoot segments and leaves are more common. Meanwhile, the morphology, physiological status, and collecting time of explants can affect the establishment of the blueberry rapid propagation system. The blueberry materials in different growth periods are collected for tissue culture. Collecting the explant materials in April is the best because the calluses induced with shoot segments are in a small quantity with a relatively high growth rate, buds differentiate more easily and browning is obviously reduced. Current-year blueberry branches with robust growth and free of pests and diseases are collected in October and cut into 3–4 cm segments for tissue culture. The corresponding plantlets obtained show a low contamination rate and strong growth.

See Fig. 6-1 for the process flow of tissue culture and rapid propagation of common blueberry explants.

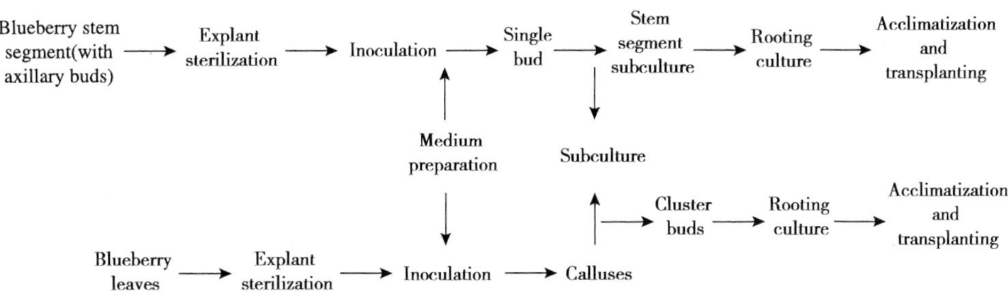


**Fig. 6-1 Process Flow of Blueberry Tissue Culture and Rapid Propagation**

## III. Primary Culture

On a sterile bench, remove both ends of sterilized tender branches with special sterilized scissors. Then cut the branches into segments about 1 cm long with 1–2 internodes and inoculate uniformly on the primary induction medium. During inoculation, 75% alcohol is often used to disinfect the inoculation tools to avoid cross-contamination. For culture conditions, the temperature is (25 ± 2)°C, the photoperiod is 14 h/d, and the lighting intensity is 2,000 lx. When axillary buds of 2–3 cm long appear, they are cut for expanded propagation.

**Knowledge Points**

Axillary buds grow in the leaf axils of plants, and they do not germinate normally. In the tissue culture, especially in the tissue culture of woody plants, stem segments are often used as explants for rapid propagation.

In the primary culture, if the explants are not thoroughly disinfected, or the culture medium is not thoroughly sterilized and operations are improper, a large number of tissue culture plantlets, or even all of them, will be contaminated. The reasons mainly include the following: The plant material itself carries microbes; It is caused by operators and microorganisms in the air; It is caused by cross-contamination resulting from the culture media and operating tools that carry microbes, and the main pathogens causing contamination are bacteria and fungi.

(1) Bacterial contamination is mainly characterized by the emergence of slimy or foam-like substances near the inoculated materials or turbid and cloudy traces on the culture medium near the materials. This usually happens 1–2 days after inoculation. Bacterial contamination is mainly caused by explants carrying bacteria or unthorough sterilization of culture media and careless operations. The solution to the bacterial contamination of explants is to sterilize the explants thoroughly: Select appropriate disinfectants and disinfection methods according to materials; Pre-treat special materials first; Add an appropriate amount of antibiotics to the material tissues carrying bacteria in the culture medium to achieve the best disinfection effects.

(2) Fungal contamination is mainly caused by molds. Generally, plaques in various colors can be found in the culture medium 3–8 days after inoculation. It is mainly caused by contaminated air, contaminated edge of the flask mouth, dust, and fungal spores falling into the flasks when the sealing film is put on the flasks or removed. Therefore, before the use of the clean bench in the inoculation room each time, the surface should be disinfected with ultraviolet light for 30 min. The culture medium should be put onto the clean bench in advance for UV surface disinfection. During inoculation, the flask should be held

obliquely with its mouth near the flame. This is to prevent spores in the air from falling into the flask, as the hot air around the mouth will rise. The sealing film should be removed gently.

In addition, the instruments and culture media used for tissue culture and inoculation should also be fully sterilized before use.

## IV. Subculture

When the axillary buds of blueberry primary culture grow up to 2–3 cm long with too many tillers or much medium loss, subculture is required: The properly sized blueberry cluster buds are cut into single buds and inoculated on fresh subculture medium for subculture. Alternatively, the flask plantlets are cut into stem segments with 1–2 leaves for subculture. After inoculation, they should be cultured in a culture room with a temperature of 25°C, lighting intensity of 2,000 lx, and a photoperiod of 12 h/d. During the subculture of tissue culture plantlets, it is important to prevent the contamination, browning, and vitrification of tissue culture plantlets.

**Knowledge Points**

Vitrification is one of the major obstacles in blueberry tissue culture. It is a phenomenon in which tender stems and leaves of some explant materials tend to be translucent and waterlogged when repeatedly propagated *in vitro*. Vitrification is a physiological lesion unique to plant tissue culture and is caused by the disturbance of plant tissue metabolism under the combined action of some physical, chemical, and biochemical factors in the culture environment. For example, the explant type, selected parts, and their sizes, temperature, lighting, sealing material, type of culture medium, and concentration of exogenous hormones will all lead to the vitrification of plantlets. ZT, agar, NAA, sucrose, and activated charcoal are all influencing factors for the vitrification of blueberry tissue culture plantlets. Reducing the concentration of cytokinin, increasing the content of agar, and adding 0.5 g/L of activated carbon in the medium can significantly reduce the vitrification of the plantlets; Increasing the amount of sucrose has no obvious effect on preventing the vitrification of the plantlets.

## V. Rooting Culture

After 20 days of subculture, the cluster buds about 2.5 cm high are cut into single buds, or blueberry flask plantlets are cut into stem segments with 1–2 leaves and transferred into the newly prepared rooting medium. The pH value is adjusted to be 5.8. After inoculation, the medium should be placed under the following culture conditions: temperature of 25°C, a lighting intensity of 2,500 lx, and a photoperiod of 12 h/d.

## VI. Acclimatization and Transplanting of Blueberry Test-tube Plantlets

To adapt the plantlets in tissue culture flasks to the external environment, acclimatization is required as follows: Add an appropriate amount of distilled water, preferably to the extent that the medium is soaked, into the flasks of plantlets before transplanting to prevent water loss and microorganism breeding in the culture medium during acclimatization. Gradually open the flasks and place them in the greenhouse for about 5 days to adapt to the environment in the greenhouse.

Take out the plantlets from the flasks and wash them to remove the medium on the roots. Soak their roots in 500–800-fold carbendazim solution and plant them in the transplanting substrate composed of sterilized garden soil, humus, and waste muscus (with a ratio of 1 : 3 : 1); then water them thoroughly, and cover them with plastic film for moisturizing. Pay attention to proper ventilation at noon every day.

## Task 2 Grape Tissue Culture

Grapes are perennial vine berry plants in the genus *Vitis* of the family Vitaceae. They are now cultivated throughout the world, ranking second only to citrus among the five major fruit trees. China ranks among the top in the world in terms of the cultivated area and yield of grapes. Because of the relatively complex genetic background of grape varieties, if traditional propagation methods are applied, the development cycle will be long. It will be difficult to cultivate new varieties, and virus infection and accumulation tend to occur, worsening the condition of virus diseases. The plant tissue culture technique is applied in the preservation, propagation, and production of grape germplasm resources.

Grapes are one of the first crops to be cultivated through tissue culture in the world. In 1961, Galzy successfully cultured grape shoot tips and performed *in vitro* conservation of grapes. In 1969, he became the first to develop the nodal segment culture method for the culture of virus-free grape shoot tips. Since then, many scholars have conducted massive research on the tissue culture of grapes, which has provided technical support for the tissue culture and rapid propagation of grape varieties. In this task, the shoot segments of a grape are used as the material for the rapid propagation and tissue culture of grape varieties.

## I. Medium Preparation

Primary induction medium: 1/2 MS+0.01 mg/L 6-BA+0.2 mg/L IAA+1.0 mg/L KT.

Subculture medium: MS+0.5–2.0 mg/L 6-BA+ 0.025–0.15 mg/L IBA.

Rooting medium: 1/2 MS+0.1–1.0 mg/L IBA.

## II. Selection and Sterilization of Explants

Collect grapes shoot and remove leaves from it to keep single-bud shoot segments 3–4 cm in length. Rinse them with tap water for 10 min; then disinfect them with 75% alcohol solution for 30 s on a sterile clean bench; rinse them again with sterilized distilled water three times; then disinfect them with 0.1% mercuric chloride solution for 6–9 min. Finally, rinse them again with sterilized distilled water 4–5 times and dry them with sterilized filter paper for inoculation.

**Knowledge Points**

1. Sterilization of Explants

Time control is very important in the sterilization of explants. If sterilization is too short or unthorough, serious contamination of one explant will affect the growth of other explants in the same tissue culture flask. If sterilization is too long, it will directly kill the explants and affect the culture process. Therefore, the appropriate length of sterilization is the key to the success of tissue culture. The appropriate disinfection length varies depending on the type of disinfectant and the selected part of the explants and generally requires specific experiments to determine.

2. Explant Materials

The explant materials used in grape tissue culture include shoot tips, axillary buds, shoot segments, leaves, petioles, and root tips. Explants all have certain effects on *in vitro* survival and bud differentiation. The apical meristem is considered to be the best choice for explant micropropagation, but there were reports that axillary buds of grapes were better for survival and regeneration than apical buds.

See Fig. 6-2 for the process flow of tissue culture and rapid propagation of common grape explants.

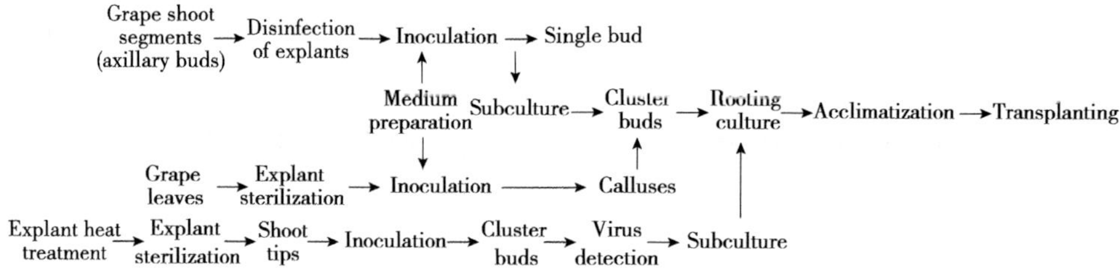


**Fig. 6-2 Process Flow of Grape Tissue Culture and Rapid Propagation**

(1) Virus-free shoot tip culture is the culture of virus-free shoot tips with the plant tissue culture technique. It is based on the fact that viruses on infected plants are not evenly distributed: Young tissues have a low content of viruses while growing points (0.1–1.0 mm) contain few or no viruses. Therefore, virus-free plantlets can be obtained through tissue culture of virus-free shoot tips. The smaller the shoot tip, the greater the chance of virus elimination, but the lower the survival rate of the explants. Generally, for a cross-cut of grape shoot tips with a length of 0.3–0.4 mm, the virus elimination rate can be higher than 60%.

(2) Heat treatment is about eliminating viruses by lowering the concentration of viruses or inactivating them in the host under a proper temperature for a suitable period of time. However, heat treatment or shoot tip culture alone cannot completely eliminate the viruses, and strict requirements

need to be met. The virus elimination rate through shoot tip culture is less than 80% and that rate through heat treatment is only higher than 26%. In contrast, the virus elimination rate through heat treatment combined with stem tip culture can be higher than 80%.

(3) Common methods for identifying virus-free grape plantlets include the indicator plant method, enzyme-linked immunosorbent assay, and electron microscopy. These methods are used to detect the grapevine fanleaf virus (GFLV), stem pitting virus (SPV), and grapevine leafroll-associated virus. When the indicator plant method is adopted, viruses transmitted through plant sap can be quickly and effectively detected using herbal indicator plants such as Gomphrena Globosa, tobacco, and cucumber. The most commonly used and most reliable indicator plants include woody plants such as *Vitis rupestris* that are sensitive to viruses. With these plants propagated by budding or greenwood grafting, viruses that are transmitted through or not through plant sap can be detected, but the detection lasts relatively long.

## III. Primary Culture

Under aseptic conditions, cut the sterilized explants into single-bud segments in a Petri dish and inoculate them into the prepared primary induction medium, with 3–4 explants inoculated in each culture flask. Place the flasks in a culture room with a temperature of (25±2)°C, a photoperiod of 14–16 h/d, and a lighting intensity of 2,000–3,000 lx. After 20 days, 2–3 cm long axillary buds will be formed.

## IV. Subculture

On a sterile clean bench, cut the primary culture plantlets into segments with 1–2 leaves with scissors, transfer them to the subculture medium, and place them in a culture room under a temperature of 25–27°C and with a photoperiod of 16 h/d and a lighting intensity of about 3,000 lx. The tissue culture plantlets can be obtained after about 28 days.

## V. Rooting Culture

When the subculture plantlets grow to about 2 cm, inoculate strong ones into the rooting medium in a culture room with a temperature of 25–28°C, a photoperiod of 14–16 h/d, and a lighting intensity of 2,000–3,000 lx. After about 30 days, the tissue culture plantlets with strong roots can be acclimatized and transplanted.

## VI. Acclimatization and Transplanting of Plantlets

When the test-tube plantlets with good root growth grow to 3–4 cm high, open the flask in the greenhouse or culture room, add a small amount of water, and open the flask gradually for acclimatization of about 5 days. Then, wash the roots with sterile water to remove the medium, transplant the plantlets into plastic basins with a substrate of peat soil and sandy loam (with a ratio of 1 ∶ 1), and cover them with plastic film. Keep the indoor temperature at 25°C and relative humidity at 100%, spray water, and ventilate every day. After 10 days of acclimatization, plantlets can be transplanted into nutrition pots filled with nutrient soil, sand, and composted organic fertilizer with a ratio of 1 ∶ 1 ∶ 0.2, and cultivated in a greenhouse.

# Task 3　Kiwifruit Tissue Culture

Kiwifruit (*Actinidia delikiosa*) is the berry of vines in the genus *Actinidia* of the family Actinidiaceae. There are 62 species and 43 varieties of kiwifruit native to China, including *Actinidia chinensis*, *Actinidia deliciosa*, *Actinidia kolomikta*, *Actinidia arguta*, and *Actinidia polygama*. Kiwifruit has a high nutritional and medicinal value. The pulp contains carotene, and vitamins C, $B_1$, $B_2$, E, and P. It also contains potassium, calcium, magnesium, ferrum, sodium, iodine, zinc, and other elements, and contains about 7.0%–22.0% soluble solids with a total acid content of 0.75%–1.95%.

Currently, the main conventional breeding methods of kiwifruit are the grafting of seedlings and cuttings of adult plants. Graftings grow slowly, with a cultivation cycle of up to two years or more, while cuttings have a low survival rate. In recent years, the cultivation area of kiwifruit in China has been expanding, and the traditional cultivation method can no longer meet the needs of large-scale production, restricting the promotion of superior cultivars. Rapid propagation with the tissue culture technique has become an important way of cultivating superior cultivars. Tissue culture is adopted for cultivation in the world's leading kiwifruit-producing countries.

In this task, the leaves of *Actinidia arguta* are used as materials to conduct tissue culture and rapid propagation of *Actinidia arguta* plantlets through callus induction.

## I. Medium Preparation

Leaf callus induction medium: MS+1.0 mg/L 6-BA+0.5 mg/L IAA+0.5 mg/L 2,4-D.

Adventitious bud differential medium: MS + 4.0 mg/L 6-BA + 0.2 mg/L IBA.

Rooting medium: 1/2 MS + 0.2 mg/L IBA.

**Knowledge Points**

The minimal media commonly used for kiwifruit tissue culture are MS and 1/2 MS, of which 1/2 MS is mainly used for rooting culture. In addition, Nitsch, N6, GD, and modified MS (nitrogen halved or nitrogen and iron salt halved) are also used. For kiwifruit tissue culture, solid media are mainly used while liquid media are seldom used. The reason is that it is not easy to control the air permeability of liquid media, thus affecting the culture effects.

The type and proportion of hormones in the culture medium are also critical factors influencing plant tissue culture. Most scholars have suggested that ZT must be used in kiwifruit tissue culture to induce regenerated plants. For the leaf callus induction and bud differentiation of kiwifruit, ZT is not essential and can be completely replaced by 6-BA with a high concentration to produce better effects. This is very important in tissue culture production and can significantly reduce the cost of kiwifruit tissue culture. Yan Jiangli et al. used 6-BA, IBA, NAA, and other growth regulators with appropriate concentrations to complete callus induction, bud differentiation, and root induction in red kiwifruit tissue culture.

## II. Selection and Sterilization of Explants

Take about 50 cm of stem segments of healthy female kiwifruit plants, put it into a greenhouse at 27°C for hydroponic culture, and change water every 3 days. After 15 days, when the buds on the stem segments grow to 10–15 cm, cut the tender segments and sterilize them with 0.1% mercuric chloride for 5–10 min, rinse them 4–5 times with sterile water, and then insert them into the proliferation medium for adventitious buds. After about 2 months, 3–5 new branches will grow from the young stems, and sterile leaves are cut off the new branches for callus induction.

**Knowledge Points**

According to the explants and culture techniques used, kiwifruit tissue culture can be classified into leaf and petiole culture, shoot tip and stem segment culture, pollen and anther culture, root and fruit culture, and embryo culture. The most widely applied methods are leaf, petiole, and stem segment culture. On the one hand, it is because materials are more convenient to obtain by using this culture method and the operation is simple and easy; on the other hand, all these kinds of explants have a high frequency of induction, and tissue culture plantlet obtained have stable trait, which is not suitable for industrialized production.

See Fig. 6-3 for the tissue culture process of commonly used kiwifruit explants.

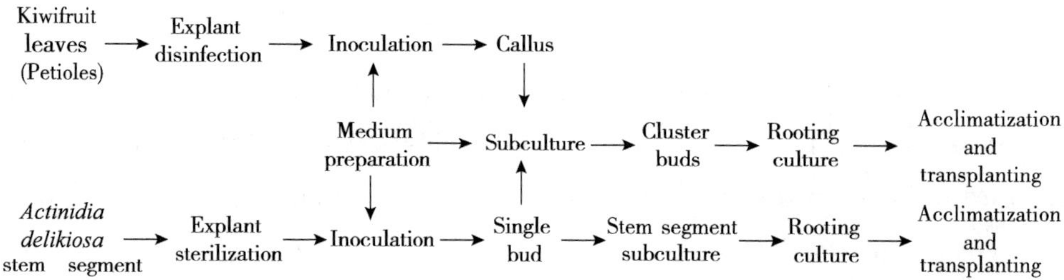


**Fig. 6-3 Process Flow of Kiwifruit Tissue Culture and Rapid Propagation**

## III. Leaf Callus Induction

On a sterile clean bench, cut off the leaf tips and margins with sterile scissors, and then cut the remaining parts into small pieces of 0.5 cm × 0.5 cm perpendicular to the leaf midrib as explants. Inoculate the leaf pieces in the leaf callus induction medium with the leaf surface facing upward. Calluses can be obtained after 30 days of culture in darkness and then transferred to the bud differentiation medium. As explants have a large cut section, attention should be paid to preventing browning during culture.

**Knowledge Points**

Browning is a common problem for plant micropropagation *in vitro*. It mostly occurs in the tissue culture of woody plants and plants with large explants. It is a process of gradual browning of the culture medium because of the release of substances from the explants during tissue culture, which eventually causes the explants to turn brown and die. The varieties, parts, shapes, sizes, selection periods, pretreatment methods, and culture conditions (such as composition, status, temperature, lighting, and pH value of the medium) of explants can all be causes of browning.

During expanded propagation in tissue culture, the following prevention and control measures for the browning may be taken.

(1) Young tissues should be selected where possible. In addition, the smaller the explant and the greater the ratio of the cut section area to volume, the greater the degree of injury and browning. Since buds do not grow easily in winter, materials in early spring and autumn are preferred as explants.

(2) Adding antioxidants, inhibitors, and adsorbents to the medium can also inhibit the enzymatic browning of the explants. Activated carbon is a strong inorganic adsorbent that can prevent browning. Polyvinylpyrrolidone (PVP) is a specific adsorbent for phenolic substances and is often used as a protective agent to prevent browning.

(3) Selecting appropriate inorganic salt composition, hormone level, sucrose concentration, and pH value during culture is essential to preventing browning. For materials that are more susceptible to browning, they should be initially cultured in darkness under a low temperature (15–20℃), which can prevent the oxidation of phenolic substances and reduce browning.

Dark culture can not only reduce browning but also facilitate morphogenesis and organogenesis. This has been proven in many plants, such as carrots, blueberries, pears, and apples. Relevant studies have shown that the dark culture of leaves has the best effects on *Actinidia latifolia*. However, chlorophyll biosynthesis requires lighting, and dark culture will affect it. As a result, the calluses formed in dark culture will be in a lighter color.

## IV. Adventitious Bud Induction

Transfer calluses in good growth to the adventitious bud induction medium for the induction of adventitious buds. The culture medium contains 3% sucrose and 0.8% agar, with a pH value of 5.8. In terms of culture conditions, the temperature is (25±2)°C, the relative air humidity is 50%–60%, the photoperiod is 16 h of light and 8 h of darkness, and the lighting intensity is 2,000–3,000 lx.

## V. Rooting Culture

Cut Regenerated shoots of 3 cm or more in height and inoculate in the rooting medium composed of 3% sucrose, and 0.8% agar with a pH value of 5.8. After 30 days, the plants with strong roots are acclimatized and transplanted.

In terms of culture conditions, the temperature is (25±2)°C, the relative air humidity is 50%–60%, the photoperiod is 16 h of light and 8 h of darkness, and the lighting intensity is 2,000–3,000 Lx.

## VI. Acclimatization and Transplanting

Before transplanting, add a small amount of water into the flask of rooted plantlets, and gradually open the flask to lower the humidity inside, allowing the plantlets to gradually adapt to the external environment. After about 7 days, take out the tissue culture plantlets, wash them with tap water to remove the medium, and then plant them in the substrate of perlite and grass charcoal (with a ratio of 1 : 4) disinfected with 0.3% potassium permanganate solution. Water them thoroughly and cover them with plastic film for moisturizing. Spray water every morning and evening and ventilate the plantlets at noon every day. Remove plastic film after 10–12 days, and then spray water every three days. Plantlets can be cultivated in greenhouses or fields after about 30 days.

## Task 4 Raspberry Tissue Culture

Raspberry (*Rubus idaeus* L.) is the fruit of a small perennial shrub belonging to the genus *Rubus* in the family Rosaceae. It is also known as Marlin and *Rubus swinhoei*. As an aggregate fruit, it has high nutritional and medicinal value and contains a variety of essential vitamins, amino acids, and minerals, especially rich in natural anti-cancer substances, SOD, and salicylic acid. It can be eaten fresh, frozen, or dried, and used for making wine, jelly, juice, and other foods, making it one of the most popular third-generation fruits in China and overseas in recent years.

Superior raspberry varieties have been introduced into China for more than 30 years, and the cultivation area has been continuously expanding. However, propagations with root suckers and cuttings are still the main breeding methods, which not only show a low multiplication coefficient rates, but also are more susceptible to seasonal in fluences. The above-mentioned deficiencies can be avoided by using the tissue culture and rapid propagation technique. Currently, in raspberry tissue culture and rapid propagation, a prominent problem is still a low multiplication coefficient that cannot meet the production needs. Tissue culture plantlets or even virus-free plantlets are mainly used for cultivation in foreign countries.

To improve the multiplication coefficient, the stem segments of the raspberry Fertod Zamatos are used for tissue culture and rapid propagation in this task.

### I. Medium Preparation

Primary culture medium: MS + 0.3 mg/L 6-BA + 0.10 mg/L NAA + 0.1% activated charcoal.

Subculture medium: MS+1.5 mg/L 6-BA+0. 3 mg/L NAA.

Rooting medium: 1/2 MS + 1.2 mg/L IBA+ 0.1% activated charcoal.

### II. Selection and Sterilization of Explants

Use the raspberry Fertod Zamatos as the test material, and select robust, disease-free, pest-free, and semi-lignified shoots with axillary buds for hydroponic culture from the production base of raspberry.

Cut the raspberry shoots into single-bud segments, wash them with detergent powder or liquid, and then rinse them with tap water for 30 min. On the clean bench, disinfect the shoot segments with 75% alcohol for 1 min, and rinse them with sterile water 3 times. Then disinfect them with 0.1% mercuric chloride for 10 min, and rinse them with sterile water 4–5 times. Finally, dry the segments with sterile filter paper for inoculation.

**Knowledge Points**

The commonly used explants for raspberry tissue culture are shoot tips, shoot segments, buds (apical, lateral, or dormant ones), and leaves. Generally, vigorous and young explants should be selected, which is related to the physicochemical properties and functions of cells in different growth periods and different parts. The growth environment, material selection, and different developmental stages of explants will lead to obvious differences in the physiological and biochemical states of explants, which will further affect the follow-up morphogenesis of tissue culture.

According to research findings:

(1) Through rapid propagation of single-bud shoot segments, the plantlets obtained are stable in hereditary characters and can be widely applied. This technique is a simple and effective method to expand and preserve precious and superior varieties.

(2) The adventitious buds formed through callus induction are remarkably variable, and repeated culture generally leads to abnormal plants. In addition, with the increase in subculture generations, the initial regeneration ability of the plantlets shown by calluses may gradually reduce or even be lost completely. To prevent plant variation, *in vitro* rapid propagation is carried out in production through the propagation of buds using buds.

See Fig. 6-4 for the process flow of tissue culture and rapid propagation of common raspberry explants.

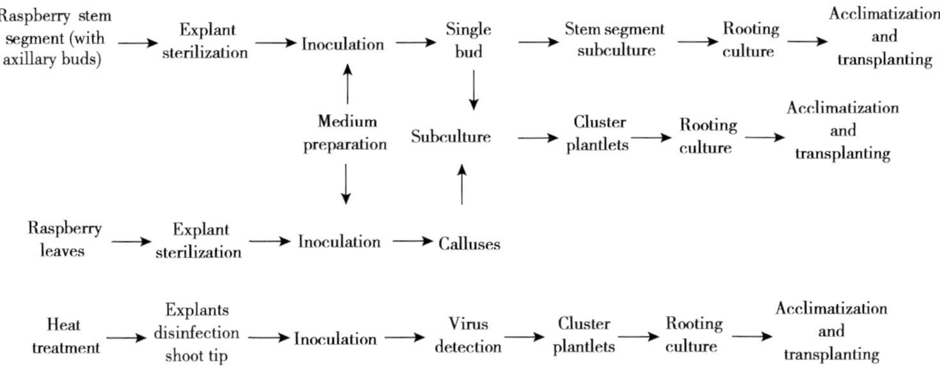


**Fig. 6-4 Process Flow of Raspberry Tissue Culture and Rapid Propagation**

## III. Primary Induction Culture

Cut off both ends of the shoot segments on a clean bench with sterile scissors and inoculate the remaining parts on the primary culture medium. The culture medium contains 3% sucrose and 0.8% agar with a pH value of 5.5. The culture conditions are as follows: Culture temperature is (25±2)°C, relative air humidity is 80%, photoperiod is 14–16 h/d, and lighting intensity is 1,800–2,000 lx.

## IV. Subculture

Cut the plantlets obtained through primary culture into segments of about 1 cm with 1–2 axillary buds and inoculate them in the proliferation medium for culture. The culture medium contains 3% sucrose and 0.8% agar with a pH value of 5.5. The culture conditions are as follows: Culture temperature is (25±2)°C, relative air humidity is 80%, photoperiod is 14–16 h/d, and lighting intensity is 1,800–2,000 lx.

**Knowledge Points**

In the tissue culture of plants, differentiation should be realized in a way that tends not to cause somatic cell variation.

(1) Use shoot segments, shoot tips, or embryoids as explants to effectively reduce variation.

(2) Shorten the subculture time and reduce subculture generations. After certain subculture generations, use new explants for culture to avoid plant degradation.

(3) Select young materials as explants.

(4) Use plant hormones of proper types and low concentrations to improve the genetic stability of plants.

## V. Rooting Culture

Inoculate the robust plantlets above 5 cm in length in the flasks into the rooting medium. The culture medium contains 3% sucrose and 0.8% agar with a pH value of 5.8. The culture conditions are as follows: Culture temperature is (25±2)°C, relative air humidity is 80%, photoperiod is 14–16 h/d, and lighting intensity is 1,800–2,000 lx. After 3–4 weeks, the plants with strong roots are acclimatized and transplanted.

## VI. Acclimatization and Transplanting

Before transplanting, transfer the culture flasks containing rooting plantlets into the greenhouse, add a small amount of water to them and gradually open the flasks to reduce the humidity inside, so that the plantlets can gradually adapt to the external environment. About 5 days later, take the tissue culture plantlets out of the flasks, wash them with tap water to remove the medium on the root system, soak them in 500–800-fold carbendazim solution, and plant them in the substrate of river sand and turfy soil (with a ratio of 1 ∶ 1). Water them thoroughly, cover them with plastic film for moisturizing, and ventilate them at noon every day. The plantlets will be ready for sale after 30 days.

# Summary

1. Tissue Culture and Rapid Propagation Methods for Fruit Trees

(1) Tissue culture and rapid propagation of blueberry: Selection of explants → disinfection of explants → primary culture → subculture → rooting culture →

acclimatization and transplanting.

(2) Tissue culture and rapid propagation of grape: Selection of explants → disinfection of explants → primary culture → subculture → rooting culture → acclimatization and transplanting.

(3) Tissue culture and rapid propagation of kiwifruit: Selection of explants → disinfection of explants → primary culture of leaves → differentiation culture→ subculture → rooting culture → acclimatization and transplanting.

(4) Tissue culture and rapid propagation of raspberry: Selection of explants → disinfection of explants → primary culture → subculture → rooting culture → acclimatization and transplanting.

2. Causes of Contamination and Preventive Measures during Plant Tissue Culture and Rapid Propagation

(1) Causes of contamination: Caused by plant materials themselves carrying microbes; Caused by the operator and microorganisms in the air; Caused by the cross-contamination of culture media and operation tools carrying microbes.

(2) Preventive measures: The solution to bacterial contamination is to sterilize the explants thoroughly. Select appropriate disinfectants and disinfection methods according to materials. Pretreat special materials first and then add an appropriate amount of antibiotics to the material tissues carrying bacteria in the culture medium to achieve the best disinfection effects.

(3) The main measures for handling fungal contamination include the following: Sterilize the clean bench and the inoculation room with an ultraviolet lamp for 30 min and then clean the bench surface with 75% alcohol each time before use. Before inoculation, wipe the exterior of flasks with 75% alcohol cotton balls and then place them on the clean bench. During inoculation, the flask should be held obliquely with its mouth near the flame. This is to prevent spores in the air from falling into the flask since hot air around the mouth rises. The sealing materials should be removed gently. The instruments and culture media used for tissue culture and inoculation should also be fully sterilized before use.

3. Identification of Grape Virus-free Plantlets

Indicator plant method, enzyme-linked immunosorbent assay, and electron microscopy. However, the most commonly used is the indicator plant method.

4. Measures to Improve the Genetic Stability of Plants

Differentiation should be realized in a way that tends not to cause somatic cell variation: Use shoot segments, shoot tips, or embryoids as explants to effectively reduce variation. Shorten subculture time and reduce subculture generations. After certain subculture generations, use new explants for culture to avoid plant degradation. Select young materials as explants. Use plant hormones of proper types and with low concentrations to improve the genetic stability of plants.

# Review Test

## I. Explain the Glossary

1. Vitrification Phenomena
2. Explant Browning
3. Virus Elimination Method for Shoot Tips
4. Virus Elimination by Heat Treatment

## II. Fill in the Blanks

1. The conventional propagation methods of fruit trees mainly include ________ and ________ .

2. Shoot tip culture for virus elimination is based on the ________ distribution of viruses on infected plants, and the growing points ( ___________ mm in area) contain little or no viruses.

3. Virus-free grape plantlets are characterized by no infection with undesirable microbes, strong adaptability, ___________ , good resistance, ___________ , strong tillering ability, ________ , mass production, ________ , and easy transportation.

4. In the plant tissue culture, fungal contamination is mainly caused by ______.

5. In the stem internodes of woody plants and some herbaceous plants, there are ________ . In general, they do not germinate and exist in dormancy, and they do not carry viruses or pathogens.

6. To remove grape viruses, ________ used in combination with ________ can significantly improve the virus elimination effects.

## III. True or False Questions

(　　) 1. Dark culture is not conducive to morphogenesis and organogenesis in plant tissue culture.

(　　) 2. Vitrification is a physiological lesion unique to plant tissue culture. It is caused by the disturbance of plant tissue metabolism under the combined action of some physical, chemical, and biochemical factors in the culture environment.

(　　) 3. For the identification of virus-free grape plantlets, herbaceous indicator plants such as Gomphrena Globosa, tobacco, and cucumber can be used to detect viruses transmitted through the sap, which is fast and very effective.

(　　) 4. Virus identification must be performed for virus-free plantlets obtained by shoot tip culture or other means.

(　　) 5. Fungal contamination is mainly caused by improper handling. Generally, plaques in various colors can be found in the culture medium 1–2 days after inoculation.

(　　) 6. According to research findings, in the virus-free culture of grapes, if the virus elimination rate of plantlets through heat treatment and shoot tip culture cannot be up to 80%, such plantlets should be discarded.

(　　) 7. During plant tissue culture, subculture time should be shortened and subculture generations reduced. New explants should be used for culture after certain subculture generations to avoid plant degeneration.

## IV. Multiple Choice Questions

1. In different parts of the same plant, _________ are found to have the lowest content of viruses.

A. Shoot tips B. Leaves C. Petioles D. Nodal cells

2. The most commonly used virus elimination method in grape tissue culture is ________.

A. Shoot tip culture B. Callus culture

C. Heat treatment D. Heat treatment combined with shoot tip culture

3. In the plant tissue culture, the concentration of alcohol used for sterilization of the sterile clean bench is ________ %.

A.50 B.75 C.95 D.100

4. In the *Actinidia arguta* callus induction culture, the leaf pieces with a size of cut perpendicular to the leaf midrib of ________ are used as explants for the test.

A.0.3 cm×0.3 cm B. 0.5 cm×0.5 cm

C. 0.7 cm×0.7 cm D. 0.9 cm×0.9 cm

5. In the plant tissue culture, the concentration of mercuric chloride used for sterilization of the sterile clean bench is ________ %.

A. 0.5 B. 0.3 C. 0.1 D. 0.05

## V. Short Answer Questions

1. Why do some fruit trees have to depend on tissue culture for rapid propagation? What are the methods of tissue culture and rapid propagation?

2. Why is variable-temperature heat treatment used in combination with shoot tip culture during the virus-free tissue culture of grapes?

3. How are vitrified plantlets formed in the plant tissue culture? What are the measures to prevent the vitrification of plantlets?

4. How to prevent contamination of plantlets during plant tissue culture?

# Module 7 Tissue Culture of Valuable Medicinal Herbs

## Main Content

This module mainly introduces the tissue culture and rapid propagation technique for medicinal plants such as *Rhizoma paridis, Dendrobium catenatum, Anoectochilus roxburghii,* and *Polygonatum cyrtonema* Hua. Through the study of this module, students will have good knowledge about the suitable medium formulas for each culture stage of *Rhizoma paridis*, *Dendrobium catenatum*, *Anoectochilus roxburghii*, and *Polygonatum cyrtonema* Hua, the process of primary culture, subculture propagation, and rooting culture of medicinal plants, as well as the methods of acclimatization and transplanting.

## Source

China is a major producer of medicinal plants. The medicinal plants here are classified into 11,146 species, in 2,309 genera of 383 families. They generally grow in the wilderness. They are also cultivated. Compared with cultivated medicinal plants, those growing in the wilderness have better quality and therapeutic effects and are favored by consumers. As a result, wild medicinal plants are consumed in large quantities, resulting in the depletion and price spikes of wild medicinal plant resources, especially for rare and endangered wild medicinal plants. Many problems exist during the cultivation of medicinal plants, such as quality degradation, pathogen-affected seeds, pesticide residues, long cultivation periods, and high cultivation costs. They may change the active ingredients and active substances of medicinal plants and largely affect the efficacy of herbal medicine.

The plant tissue culture technique boasts a short reproductive cycle, low cost, and avoidance of pesticide and heavy metal pollution. It has been increasingly widely applied to many medicinal plants and plays an important role in the conservation and sustainable use of resources. According to incomplete statistics, among more than 400 medicinal plants that have been studied for plant cell culture, over 200 medicinal plants have successfully yielded medicines by plant tissue culture.

# Work Tasks

## Task 1 Tissue Culture of *Rhizoma paridis*

*Rhizoma paridis*, a traditional Chinese medicine (TCM), is the dried root and rhizome of *Parispolyphylla* Smith var. *yunnanensis* (Franch.) Hand.-Mazz. or *Paris polyphylla* Smith var. *chinensis* (Franch.) Hara, both belong to the family Liliaceae . As the main raw materials for such products as Yunnan Baiyao series and Gongxuening Capsules, its medication has been detailed in *Shen-nong's Herbal Classics* and *Compendium of Materia Medica*. The demand for *Rhizoma paridis* has been increasing in recent years, but its protection and breeding have been neglected. Over-exploitation of wild resources, being a predominant issue, was caught in a vicious cycle, resulting in a serious overdraft of wild resources. Therefore, it is urgent to upgrade artificial planting and breeding, of which plantlet breeding is the source of artificial cultivation research of wild resources, standardized planting of TCM materials, and development of the entire TCM industry. Thus, the tissue culture technique is one of the ways to solve the bottleneck in the asexual rapid breeding of *Rhizoma paridis*, which has certain practical significance.

## I. Selection and Sterilization of Explants

The effects of different parts of *Rhizoma paridis* vary greatly as explants to induce callus. It has been found that apical buds and rhizomes, except for aboveground parts such as leaves, sepals, stem segments, and petioles that are not suitable for tissue culture, can induce callus as explants.

First, select the plump seeds, remove the glumes with tap water, rub the seed coats by hands, and screen out the seed coats of red berries with a fine sieve to remain white seeds. Then, wash the seeds with sterile water 3–5 times on the clean bench, soak them in a beaker containing 75% alcohol for 30 s, and gently shake. After that, wash with sterile water for 3–5 times, soak the seed explants with 0.1% mercuric chloride for 10–15 min (continuously shake the beaker to ensure that all seeds come in contact with the solution evenly), rinse with sterile water for 3–5 times. Finally, dry the seed surface with sterile filter paper, and inoculate on the callus-inducing medium (MS + 0.5 mg/L 6-BA + 0.1 mg/L NAA).

Take the apical buds of *Rhizoma paridis*, rinse them with tap water for 2–3 h, remove the coleoptiles on the surface, rinse with tap water, and then rinse with distilled water 3 times. Under aseptic conditions, disinfect with 75% alcohol for 30 s, rinse with sterile water 3 times, sterilize with 0.1% mercuric chloride for 15 min, and then rinse with sterile water 5 times. Peel off the coleoptiles outside the bud in layers, cut into small pieces with a length and width of about 1 cm respectively, cut the non-layered parts inside the bud into small pieces according to the size of the coleoptile, and then inoculate on the callus-inducing medium (MS + 2mg/L 6-BA + 0.1 mg/L NAA).

Flush away the surface sediments of rhizomes of *Rhizoma paridis* collected in the field with running water, remove the lateral roots. Then, further brush off the surface sediments with a banister brush under running water, and cut off the apical buds at 1.5–2.0 cm downward from the blastema with a sharp and clean blade for later use. After rinsing with running water, routinely disinfect and sterilize the surface on the clean bench, sterilize with 75% alcohol for 15 s, rinse with sterile water 3 times, then sterilize with 0.1% mercuric chloride for 8 min, rinse with sterile water for 5 times, peel off one to two layers of coleoptiles, and retain 1 cm downward from the blastema for inoculation on the medium (MS+1.0mg/L 6-BA +1.0 mg/L NAA).

**Knowledge Points**

Currently, *Rhizoma paridis* has a low frequency of callus induction. The callus exhibits a granular appearance, loose texture and easy dispersion. If the culture lasts too long, the plant will be susceptible to browning, fibrosis, and loss of propagation and differentiation capacities. Therefore, in addition to the selection and disinfection of explants and the screening of media during tissue culture, it may also be influenced by such factors as low ability of tissue cell dedifferentiation and slow cell division of *Paris* L., as well as the ecological environment and autogenetic inherited characteristics of *Paris* L.

## II. Primary Culture

Culture in a thermostatic room with a temperature of 20–25°C, a lighting intensity of 2000 lx, and an illumination length of 12 h/d.

The callus induced by seeds is relatively loose, highly humid, and fragile. However, the callus induced by seeds will not differentiate into plantlets.

After the pieces of apical buds are inoculated on the callus-inducing medium for 25 days, the inoculated pieces will begin to expand, and about 15 days later, a yellowish callus with a rough surface and a hard texture will gradually break out at the incision of the inoculated piece.

The apical buds begin to expand after inoculation of the rhizome of *Rhizoma paridis* for 21 days. Tubercules begin to appear at the base of apical buds after 27 days, and granular protrusion appears in some coleoptiles of apical buds.

## III. Subculture

Transfer the cutting piece of induced callus to the propagation and differential medium (MS+2 mg/L 6-BA +0.5 mg/L NAA +0.5 mg/L KT), and then the callus will begin to proliferate slowly. After 90 days, transfer it to a new medium. After about 180 days, the callus piece proliferated to about 0.5 cm in diameter and gradually changes from yellowish to white, and the surface gradually becomes smooth from rough protrusion. After another 30 days, it gradually differentiates to form a bud. If the differentiated buds are not induced for rooting, they will continue to grow on the medium for about 150 days to spread leaves and form intact rootless plantlets. A proliferation cycle lasts for 210–240 days, and adventitious buds can propagate 1–2 times in each cycle.

### IV. Rooting Culture

Cut off the differentiated bud and inoculate on medium (1/2 MS+0.5 mg/L NAA +0.5 mg/L IAA) for rooting induction. During the culture process, the base of the bud gradually browns and elongates into a rhizome. After culture for about 60 days, 2–3 roots will grow at the base of the brown and elongated front-end bud, with a rooting rate of 76.3%.

### V. Acclimatization and Transplanting

Remove the tissue culture plantlets from the culture room with flasks, acclimatize the plantlets in a greenhouse, gradually open the flask caps after 15 days, and then spray water once every morning, noon and evening to ensure adequate humidity. Gently pick out the plantlets with tweezers, wash out the root medium, transplant them in the substrate with loam and leaf mould (with a ratio of 1 : 1) at a temperature of 18–20°C, and keep the soil humidity at 50%–60%.

## Task 2 Tissue Culture of *Anoectochilus roxburghii*

*Anoectochilus roxburghii* is a perennial herb of Orchidaceae, also known as *Anoectochilus roxburghii* (Wall.) Lindl. and *Anoectochilus taiwanensis* Hayata, mainly distributed in Fujian, Guangdong, Guangxi and Yunnan. At present, *Anoectochilus roxburghii* available on the market mainly comes from Fujian (*Anoectochilus roxburghii*), Taiwan (*Anoectochilus formosanus* Hayata), Yunnan and Vietnam [ *Anoectochilus roxburghii* (Wall.) Lindl. ]. As they are in the same genus and have similar morphology, the above three plants are used as *Anoectochilus roxburghii* in folk medicine. *Anoectochilus roxburghii* is a precious traditional Chinese medicinal herb. With all parts being medicinal, it can treat diseases such as cough with lung heat, rheumatism and bone pain, diabetes and cystitis by removing heat to cool blood, dehumidifying and detoxifying, thereby promoting longevity. *Anoectochilus roxburghii* is small and beautiful, with a high ornamental value and a broad application prospect. The seeds of *Anoectochilus roxburghii* are small, with a very low germination rate under natural conditions. The plants are short and slow in growth and development. They are favored by insects and birds, making them more valuable and rare. Due to the slow propagation of *Anoectochilus roxburghii* from cuttings, the industrialized plantlet is mainly realized by tissue culture, so as to create conditions for conservation, reasonable development and utilization of wild resources. Currently, the tissue culture technique of *Anoectochilus roxburghii* is relatively mature and mainly realized in the following four ways: ①Induction of calluses through dedifferentiation of explants followed by inducing the differentiation of buds; ②Direct induction of the explants to germinate axillary buds or apical buds, followed by subculture; ③Induction of seeds to germinate into protocorms to obtain bud plantlets through propagation and differentiation; ④"One-step plantlet regeneration" tissue culture by optimizing the media and culture conditions.

## I. Selection and Sterilization of Explants

Select the indehiscent mature capsule, rinse with tap water, wipe the pericarp with a 75% alcohol pad, soak it in 10% sodium hypochlorite solution for 10–12 min, and rinse with sterile water 5–6 times.

Select vigorously-grown plant of *Anoectochilus roxburghii* that is free from diseases and pests, and pick the segments and adventitious buds as the explants. Brush off the surface with a banister brush, place it in a small beaker and cover it with gauze. Rinse it with running water for about 30 min, then soak it in 75% alcohol for 30 s on the clean bench, sterilize it with 0.1% mercuric chloride for 8 min, and rinse with sterile water for 5 times.

**Knowledge Points**

Numerous studies have shown that structured plant tissues or organs can produce callus when cultured *in vitro* under certain conditions. Among the explants such as leaves, stem slices, stem segments with nodes and adventitious buds of *Anoectochilus roxburghii*, the frequency of callus induction of adventitious buds is the highest, while the leaves are not suitable as explants for callus induction. The seeds of *Anoectochilus roxburghii* are small and look like dust. A capsule contains thousands of seeds that consist of seed coats composed of globular embryos and monolayer cells. Under natural conditions, the seeds can germinate and grow through symbiosis with fungi.

## II. Primary Culture

Longitudinally dissect the disinfected capsule into two pieces with a dissecting needle, pick a few seeds with tweezers, and sprinkle them into the medium (1/2 MS+ 1.0 mg/L 6-BA+1.0 mg/L NAA). The embryos expand significantly after culture for about 1 month, break through the seed coats to germinate, and can develop into protocorms after culture for about 45 days. Inoculate the protocorms on the medium (MS + 0.2 mg/L 6-BA +0.5 mg/L NAA) for callus induction, and the frequency of callus induction can reach 52.5%. After the callus gradually differentiates into protocorm-like bodies (PLBs), inoculate PLBs on the differential medium (MS + 1.0 mg/L 6-BA + 0.5 mg/L NAA). The base of PLBs can further differentiate into callus and new PLBs.

Cut the segments into smaller segments (about 1 cm), and culture in medium ( MS+0.5 mg/L NAA + 5.0 mg/L 6-BA +0.3 mg/L KT ) under aseptic conditions.

Cut off the adventitious bud, and inoculate on medium ( MS+0.5 mg/L NAA + 4 mg/L BA +0.3 mg/L KT ) for culture. After inoculation for 13 days, the adventitious bud begins to expand, with small tubercules that are milky yellow and shiny. After inoculation for 25 days, the tubercules continue to increase, forming a large piece of callus that is bluish-green, with globular milky protrusions on the top. The frequency of callus induction of adventitious buds is up to 96.67%.

Inoculate the callus piece of the adventitious bud on the differential medium (MS+0.5 mg/L NAA + 2 mg/L 6-BA) for culture, and brownish-red bud spots will appear at the edge of the callus piece and grow into small buds on Day 16, which continue to grow and develop into plantlets.

Culture in a culture room, with a temperature of 26–28°C, a lighting intensity of 1,500–2,000 lx, and a photoperiod of 10 h/d.

**Knowledge Points**

Protocorms can directly develop into plantlets and also produce callus that then develops into PLBs and differentiates into plantlets. PLBs can proliferate without cutting, and some of them gradually differentiate into buds, while some are in a meristematic state, but they can eventually form plants by subculture. When the protocorm is transferred to the induction medium, its top will differentiate into leaves, while a large number of white rhizoids will grow on its base. The rhizoids can extend into the medium to absorb nutrients. The rhizoids remain for a considerable period of time, and then gradually disintegrate after root differentiation.

## III. Subculture

Inoculate PLBs on proliferation medium (MS + 1.0 mg/L 6-BA + 0.1 mg/L NAA) for proliferation culture, and the proliferation rate can reach 6–7 times after culture for 60 days.

The optimal medium for the induction and proliferation of cluster buds of *Anoectochilus roxburghii* is MS + 2.0 mg/L 6-BA + 1.0 mg/L NAA + 0.1 mg/L KT + 20% coconut juice + 30 g/L sucrose + 7 g/L agar (pH 5.8).

## IV. Rooting Culture

Spread the *in vitro* stem segments of *Anoectochilus roxburghii* flat and inoculate on the medium (1/2 MS + 0.5 mg/L 6-BA + 1.0 mg/L NAA + 100 g/L banana juice + 1.0 g/L activated carbon + 20 g/L sucrose). After inoculation for 10 days, the nodes begin to germinate and protuberate to form milky and granular aerial root primordia, which subsequently gradually elongate to aerial roots. Simultaneously, the axillary bud primordia on the nodes begin to germinate to form bud tips, which gradually grow upwards to form intact plants with stems, leaves, nodes and roots (including aerial roots and basal roots). When cultured for 90 days, the rooting rate reaches 100%.

The optimal rooting medium for strong plantlets of Fujian *Anoectochilus roxburghii* is 1/2 MS + 0.1 mg/L 6-BA + 0.1mg/L IBA + 20% coconut juice + 20% mashed potato + 30 g/L sucrose + 6.5 g/L agar.

The optimal rooting medium for strong plantlets of Taiwan *Anoectochilus roxburghii* is 1/2 MS + 0.1 mg/L 6-BA + 0.5 mg/L IBA + 20% coconut juice + 20% mashed potato + 30 g/L sucrose + 6.5 g/L agar.

**Knowledge Points**

Buds in tissue culture can absorb the nutrients from the medium to grow normally even without roots, but they are difficult to survive when transplanted to the field. Additionally, the tissue culture plantlets are generally thin and weak, so a strong plantlet and rooting culture are basically required before transplanting.

## V. Acclimatization and Transplanting

Before transplanting, acclimatize the tissue culture plantlets of *Anoectochilus roxburghii* in a cool and ventilated place for 1–2 weeks. Then, wash out the agar medium adhered to the rhizomes, and disinfect with 400-fold-diluted carbendazim for 10 min. After that, transplant into humus soil, with a planting density of 3 cm × 3 cm and a shading rate of 80%, and spray once a day after transplanting.

**Knowledge Points**

The cultivation of *Anoectochilus roxburghii* in greenhouses can be done with both plastic baskets and shelves. For cultivation in plastic baskets: Adopt a 45 cm × 40 cm plastic basket, place gauze element and lay with substrates of about 10 cm in thickness, and then plant *Anoectochilus roxburghii* in humus soil, with 100 plants per basket. For cultivation on shelves: Erect wooden beds (length × width × height=1.5 m × 1 m × 1 m) in the greenhouse, and spread forest humus soil (about 10 cm in thickness) flat on the bed. The survival rate of *Anoectochilus roxburghii* can reach 90%–92%. Some studies have found that the habitat conditions of *Phyllostachys pubescens* forest and hardwood forest are close to the wild state, and are more suitable for the growth of *Anoectochilus roxburghii*.

## Task 3 Tissue Culture of *Dendrobium catenatum*

*Dendrobium nobile* is a valuable TCM in China, which is divided into dozens of varieties such as *Dendrobium catenatum*, *Dendrobium chrysanthum, Dendrobium nobile* Lindl., *Dendrobium huoshanense, Dendrobium loddigesii* Rolfe and *Dendrobium fimhriatum* Hook.var.*oculatum* Hook., of which *Dendrobium catenatum* (its dry product also known as *Dendrobium officinale*) is the most valuable and of the highest grade, and ranks first in the "Nine Precious Chinese Immortal Herbs". As the old Chinese saying goes: "Ginseng in the North and *Dendrobium officinale* in the South", and it is also called "life-saving immortal herb" "Chinese immortal herb" "gold in medicine" in folk. It is mainly produced in Yunnan, Guizhou, Zhejiang, Sichuan, Guangxi, and Anhui, mostly growing in tropical and subtropical mountains, cliffs, and rock cracks with high temperatures and humidity. It is an epiphytic orchid with slow growth and very low reproductive capacity. It is also very delicate and will die from hard light, direct light, or rainstorm and snow freezing. Excessive moisture will also rot its roots and even kill the whole plant. The great medicinal value of *Dendrobium catenatum* leads to long-term unrestrained collection, making this plant threatened with extinction. Therefore, it is classified as a National Second-level Protected Plant in China. The tissue culture and rapid propagation technique is an effective way to solve the shortage of wild resources and protect the germplasm resources of *Dendrobium catenatum.* Establishing a stable and effective tissue culture and rapid propagation system and cultivation technique system can not only provide a large number of plantlets for large-scale artificial cultivation of *Dendrobium catenatum*, and meet the market demand for *Dendrobium* medicinal resources, but also increase the population quantity of *Dendrobium catenatum*, improve its endangered status, and provide a strong guarantee for the protection and sustainable use of germplasm resources.

## I. Establishment of Asepsis System

### 1. Aseptic Sowing

Select the indehiscent capsule of *Dendrobium catenatum*, rinse with running water for 0.5–1 h, wipe the surface with 75% alcohol. Then, process with 0.2% mercuric chloride for 15 min, and rinse with sterile water 3 times. After that, cut the capsule on sterile filter paper, and sow the seeds on the seed germination medium

(MS+0.5mg/L NAA+ 10% potato juice).

2. Culture of Stem Segments

Select young branches of vigorously-grown *Dendrobium catenatum*, peel off the leaves and membranous leaf sheaths on the nodes, wash with detergent, and gently scrap off the surface dirt with a scalpel. Rinse with tap water for 30 min. Cut the branches into about 6 cm stem segments on the clean bench, disinfect with 75% alcohol for 5 min, and rinse with sterile water 3 times. Then, disinfect with 0.1% mercuric chloride solution for 10 min, rinse with sterile water for 3 times, and soak the stem segments with sterile water for 30 min. Continuously and gently shake the beaker during disinfection to make the stem segments come in full contact with the disinfectant. Dry the surface of disinfected stem segments with sterile filter paper, cut them into 1.5 cm segments with nodes, and inoculate on the induction medium of adventitious buds and protocorms.

**Knowledge Points**

Currently, the explants of *Dendrobium catenatum* mainly include seeds, stem segments, shoot tips, protocorms and artificial seeds. Studies showed that the callus induced by seeds has better differentiation ability, and the quality of protocorms cultured by seed embryos is also higher. Currently, the main ways to obtain regenerated plants include seeds → protocorms → plantlets; seeds → protocorms → stem segments of aseptic plantlets → plantlets; seeds → protocorms → calluses → cluster buds → rooting plantlets; seeds → calluses → protocorms → plantlets; protocorms → artificial seeds → plantlets; stem tips → calluses → cluster buds → rooting plantlets.

## II. Primary Culture

The temperature of the culture room is 25±2°C, the photoperiod is 12 h/d, and the lighting intensity is 2,000 lx.

After seeds are cultured for about 10 days, the proembryo gradually germinates and expands. About 20 days later, one end of the seed coat ruptures, and the embryo breaks through the seed coat in a small conical shape, i.e. the protocorm. Afterwards, the top of the protocorm protrudes from the original semi-concave shape to form scale-like leaf primordia that then grow into 1.0–1.5 cm plantlets after about 3 months.

The stem explants are inoculated on the protocorm induction medium (MS+1.0 mg/L 6-BA +1.0 mg/L IBA + 60 g/L mashed potato + 0.5 g/L activated carbon + 25 g/L sucrose + 6.5 g/L agar). After inoculation for 45 days, the protocorm inductivity is 97.78%.

Stem explants are inoculated on the medium (MS+1.0 mg/L 6-BA +0.5 mg/L NAA). After inoculation for 10 days, the buds in their stem nodes protrude and generate cluster buds that grow rapidly and stoutly into robust and complete buds after 30 days, without induction of protocorms.

## III. Propagation and Differentiation Culture

The propagation of test-tube plantlets is critical for plant tissue culture. Thc protocorms are induced to generate cluster buds that are then cultured into cluster plantlets, so as to achieve rapid propagation of plantlets. The protocorms of *Dendrobium catenatum* are loose and in a mulberry-like shape, and their quantity and quality are extremely important for proliferation culture. When the protocorms develop into cluster buds, if there is no suitable medium, the morphogenesis of plantlets and the quality of test-tube plantlets will be affected, resulting in malformed plantlets and affecting the expansion and propagation of test-tube plantlets.

1. Proliferation Culture of Protocorm

Separate the protocorms obtained from the primary culture into small clusters with tweezers, make the differentiation tips upward, and inoculate them on the proliferation medium (MS+2.5 mg/L 6-BA +1.0 mg/L IBA). The PLBs particles through proliferation culture are full and uniform in growth, and the differentiated leaf primordia protrusions on a large number of PLBs can be observed, with a proliferation factor of 7.94 after culture for 30 days.

2. Differentiation Culture of Protocorm

Disperse the proliferatively cultured protocorms, and inoculate the well-grown PLBs on the differential medium (MS+2.5 mg/L 6-BA +1.0 mg/L IBA) to culture for 60 days. The inductivity of cluster buds is 97.33%.

3. Proliferation Culture of Cluster Buds

Cut the cluster buds induced by stem explants into single buds or small bud clusters. Then, transfer and inoculate on the proliferation medium (MS+1.0 mg/L 6-BA+0.5 mg/L NAA+ 100 g/L potato juice) as per 15 buds/flask, and culture for about 30 days as a cycle, with a proliferation coefficient of 7. The average proliferation coefficient is small for the breeding route with stem segments as explants, but the proliferation cycle is short. The leaves are dark-green and the plantlets grow vigorously. They can be directly used for rooting without strong plantlets.

## IV. Rooting Culture

After the proliferation subculture, some flask plantlets of *Dendrobium catenatum* have 1–3 roots, which are underdeveloped, and some flask plantlets have no roots. Therefore, when the subcultured cluster buds of *Dendrobium catenatum* are inoculated on the rooting medium (1/2 MS + 2.5 mg/L NAA + 1.5 mg/L IBA +150 g/L banana puree), the rooting rate can reach 100%. There are white powders on the roots, and flask plantlets are of bright leaves, thick stems and many nodes, which are conducive to transplanting.

## V. Acclimatization and Transplanting

Transplant bagged plantlets in cool seasons (March – May or October – November), and acclimatize plantlets for about 15 days before transplanting. After removing the bagged plantlets, thoroughly rinse the agar at the roots with clear water, and dry them for planting. Select the pulverized granules of perennial Korean pine bark as the cultivation substrates, kill pathogens and eggs by autoclave, and conduct deresination. Then, soak in water for 24 h, and spread on the plantlet tray at a thickness of 3 cm. After that, provisionally plant the plantlets on the plantlet tray at a planting spacing of 1 cm × 2 cm for 3 months, protected from light by 95%, water once every 3–7 days, and apply 10–15 g of slow-release fertilizers. After 3 months, transplant the robust grade I plantlets to the seedbed covered with 5–8 cm thick and disinfected pulverized granules of Korean pine bark, at a planting spacing of 10 cm × 15 cm. After stable growth, change the shading rate to 75%–80%, water once every 3–5 days, apply the slow-release fertilizer and self-prepared 2‰ liquid manure [dissolve 2,470 mg of Ca $(NO_3)$, 800 mg of $KNO_3$, 450 mg of $MgSO_4$, and 200 mg of $KH_2NO_3$ per liter of water] in combination, and spray once every 10 days.

## Task 4 Tissue Culture of *Polygonatum cyrtonema* Hua

*Polygonatum cyrtonema* Hua, a perennial herb belonging to the *Polygonatum* of Liliaceae, grows under trees, in scrubs, or in the shade of hillsides. Its rhizomes are fleshy and can be used as food and medicine, with the effects of boosting qi and nourishing yin, invigorating spleen, moistening lung, and tonifying the kidney. It shows potential medicinal value in anti-aging, immunity adjustment, blood lipid and blood glucose regulation, memory improvement, and anti-tumor and anti-bacterial immunity enhancement, with promising prospects of development. However, the wild plant resources of *Polygonatum* have been endangered, and artificial cultivation is only in the initial stage. Artificial cultivation is mainly realized through tuber and seed propagation. Tuber propagation poses problems such as the consumption of medicinal material resources and variety degeneration, while seed propagation faces problems such as dormancy, low germination rate, and long plantlet raising cycle. Therefore, it is impossible to meet the market demand for *Polygonatum cyrtonema* Hua through natural propagation. Compared with traditional cultivation methods, tissue culture is not restricted by such factors as season and number of materials and can produce high-quality plantlets with stable properties on a large scale in a short period of time.

### I. Sterilization of Explants

Select full seeds not affected by diseases or pests, add two drops of detergent, and rinse them under running water for 2–3 h. Soak them with 75% alcohol for 20 s on the sterile clean bench, and rinse with sterile water 2–3 times. Then, disinfect with 0.1% mercuric chloride solution for 5 min, and rinse with sterile water 4–5 times.

Select high-quality budded tubers of *Polygonatum cyrtonema* Hua. Gently brush off the surface soil under running water. Then, remove the leaves and roots, cut off the buds from the rhizomes, and take 2–3 cm apical buds. After that, put them into flasks, rinse with running water for 1–2 h, soak in 75% alcohol for 15 s on the clean bench after drying. Rinse with sterile water 2–4 times, put them into 0.1% mercuric chloride solution and shake continuously. Finally, take them out after 8–10 min, and rinse with sterile water 5 times.

## II. Primary Culture

Inoculate sterilized seeds on medium (MS + 2. 0 mg /L 6-BA).

Inoculate sterilized budded tubers on the induction medium (MS + 2. 0 mg /L 6-BA + 1. 0 mg/L KT + 0. 2 mg/L NAA + 0. 5 mg/L 2, 4-D), under a temperature of (25 ± 1)°C, a photoperiod of 14 h/d, and a lighting intensity of 2,000–3,000 lx.

## III. Subculture

When seeds have stem leaves and radicles, cut a segment of stem and transfer it into the subculture medium (MS + 1. 0 mg/L 6-BA + 0. 5 mg/L NAA +0.3 mg/L IAA). One week later, after new buds appear, continue the propagation of new buds.

Cut tubers with differentiated adventitious buds into small cubes (0.5 $cm^3$) and transfer them to the proliferation medium (MS + 4. 0 mg /L 6-BA + 0.2 mg/L NAA). The proliferation factor can be 5.67.

## IV. Rooting Culture

Transfer the buds (3.0–3.5 cm) to the rooting medium [ 1/2 MS + 1. 0 mg/L NAA+ 0.2 mg/L CA (activated carbon) ] for culture. The rooting rate can reach 96.8%, with an average rooting number of 22.7, and the root system is stout with vigorous growth.

## V. Acclimatization and Transplanting

Select tissue culture plantlets with robust roots for acclimatization and transplanting, and uncap culture flasks and acclimatize for 3 days. Select plantlets with more than three robust roots from the culture flasks with tweezers.

Gently wash the plantlets to remove the medium attached to the roots with a large amount of clear water. Soak the plantlets in carbendazim solution for 5 min and then in rooting solution for 0.5 h. After that, transplant them into substrates (peat soil ∶ sand = 2 ∶ 1).

## Summary

The tissue culture of medicinal plants breaks through the traditional propagation of medicinal plants and opens a new chapter for the asexual propagation of medicinal plants. With the tissue culture technique, medicinal plants that can be rarely sexually propagated or cannot be sexually propagated at all will be normally propagated without being limited by time and space. In addition, medicinal plantlets can be provided for thc standard mass production of medicinal plants to prevent variety degradation and optimize varieties, preserve the germplasm resources for endangered medicinal plants, prevent species extinction, and protect biodiversity.

This module introduces the tissue culture and rapid propagation techniques for *Rhizoma paridis, Dendrobium catenatum, Anoectochilus roxburghii*, and *Polygonatum cyrtonema* Hua. As shown by the examples of tissue culture and rapid propagation of these medicinal plants, the tissue culture and rapid propagation of plants are largely the same, and the key lies in the acquisition of sterile materials and the establishment of a rapid propagation system. Whether tissue culture is difficult or easy, succeed or fail, largely depends on factors such as genotype, species, source, size, collection season, physiological status, and age of explants. For *Rhizoma paridis* and *Polygonatum cyrtonema* Hua, young rhizomes with buds are better explants, while seeds or stem segments with buds are better explants for *Anoectochilus roxburghii* and *Dendrobium catenatum*.

The tissue culture of medicinal plants is mainly realized in two ways, i.e. differentiation of calluses into plantlets, and induction of apical and axillary buds into plantlets. For the former, the procedures are sterilizing explants, inoculating them on the de-differential medium to induce the calluses, transferring them to the bud、inducing differential medium to differentiate into cluster buds, continuing the proliferation subculture of cluster buds, transferring them to the rooting medium for forming plantlets, acclimatizing and then transplanting the plantlets. For the latter, the procedures are taking stem segments with buds from shoots of mother plants, sterilizing them to obtain sterile materials, transferring them to a subculture medium to form cluster buds, cutting 2–3 cm high buds, and transferring them to rooting medium to form plantlets. For the proliferation subculture of cluster buds, the regeneration process from buds to plantlets can generally be repeated after about 30 days. In this way, numerous rootless plantlets (cluster buds) can be quickly regenerated within a short time.

## Review Test

1. Briefly describe the selection of explants for tissue culture of *Rhizoma paridis*.

2. In the subculture transfer of *Dendrobium catenatum*, how to divide the cultured materials?

3. Briefly describe the process of tissue culture and rapid propagation for *Dendrobium catenatum*.

4. How to sterilize the explants of *Anoectochilus roxburghii*?

5. What are the problems worth noting during the tissue culture and rapid propagation of *Rhizoma paridis*?

6. What are the characteristics of the tissue culture and rapid propagation of medicinal plants?

7. What are the commonly used explants for tissue culture and rapid propagation of *Anoectochilus roxburghii*?

8. What are the technical points for acclimatizing and transplanting tissue culture plantlets of medicinal plants?

9. Write down a tissue culture and rapid propagation technique for medicinal plants by consulting data.

# Module 8 Management of Industrialized Production of Tissue Culture Plantlets

## Work Tasks

### Task 1 Development and Implementation of Production Plan

The industrialized production of tissue culture plantlets is about the high-efficiency production of high-quality plantlets in batches and by plan using standard, mechanical, and automation technologies under the optimal artificially controlled environmental conditions by making full use of natural and social resources. It is mainly used in plant rapid propagation and virus-free plantlet production. Currently, this technique has been adopted for the mass industrialized production of tissue culture plantlets for many cash crops such as flowers, fruit trees, and vegetables.

The industrialized production of tissue culture plantlets requires certain infrastructure and equipment. It is necessary to optimize the production process flow, make the production plan as per the market demands, and strengthen the scientific management in the production and sales process, so as to minimize production costs and obtain good economic benefits.

#### I. Formulation of Production Plan

##### (I) Formulation Principles

Formulating a production plan is the key and important basis for the commercial production of tissue culture plantlets as underproduction or overproduction will both cause direct economic losses. The general principles for formulating the production plan are to determine the annual sales goal based on market demand and trend, as well as production conditions, scale, and strength, and then comprehensively consider the losses in each procedure of tissue culture plantlet production. Finally formulate the corresponding production plan.

## (II) Formulation Basis

1. Market Survey Conclusions

Market survey conclusions are drawn based on market survey, and scientific statistical analysis and prediction. They are important in decision-making for enterprises to formulate production plans and implement scientific and effective operation and management under the conditions of the market economy. Therefore, for the commercialized production of tissue culture plantlets, a market survey should also be conducted carefully. Some tissue culture factories may face a serious backlog because of poor sales, causing huge waste and great economic losses. This is exactly where poor market survey leads due to blind production without considering market conditions and demand.

The market survey on tissue culture plantlets mainly involves the survey on market demand and market share, as well as scientific analysis and prediction of these factors. Generally, the market demand is predicted based on the regional planting structure, natural climate, plant species, and market trends. For example, potatoes are planted in North China, Northeast China, and the northern part of East China, in a large area to meet the high demand for plantlets. Herbaceous flower plantlets are in high demand in fresh-cut flower production bases in Kunming, Shanghai, and Shandong. South China shows obvious advantages in the propagation of plantlets of herbaceous flowers and ornamental forest trees. North China rises unexpectedly in the propagation of plantlets of bulb flowers, occupying an increasingly large market share in China. The so-called market share refers to the percentage of the sales volume or sales amount of a product from an enterprise to the total sales volume or sales amount on the whole market. The market share of a certain tissue culture plantlet is analyzed and predicted through the survey on its variety, quality, price, production, marketing channel, packaging, freshness, transportation mode, and advertisement. Generally speaking, if the plantlets produced by enterprises are in a dominant position in terms of quality, price, supply time, and packaging, sales volume and market share will be large, and vice versa.

The scientific conclusion is drawn based on the market survey, analysis, and prediction of tissue culture plantlets, and is used as a guide to formulate the production plan for tissue culture plantlets. Only in this way can targeted production of tissue culture plantlets be carried out to avoid blind production.

2. Supply Quantity and Supply Time

If there are stable orders, the production can be organized based on order quantity for delivery on schedule. If the lead time is relatively long and plantlets need to be supplied in stages and batches from autumn to next spring, proliferation and rooting induction may be carried out at the same time after the $4^{th}$ to $5^{th}$ subculture, since the best proliferation effects can be achieved during the $4^{th}$ to $10^{th}$ subculture. If the supply is concentrated in a certain period and there is enough time for subculture, continuous proliferation may be carried out until there are enough plantlets, and then plantlet growth and rooting may be promoted at a time before plantlets are supplied. If orders are received late, i.e. the lead time is short, it is often necessary to increase the plantlet base number by increasing the proliferation coefficient in the early stage (hormones may be used for regulation, the proportion of cytokinins may be improved in particular and optimal temperature and photoperiod should be maintained). If there are no large orders, the number of plantlets to be proliferated must be restricted and the proliferation and growth rate must be consciously controlled. Generally, the control measures include appropriately cooling, adding growth inhibitors into the medium, lowering the hormone level, and storing the original plant material at low or ultra-low temperatures. For tissue culture plantlets, especially those in flasks or being acclimatized for direct sales, once the production quantity is determined based on the market prediction, the time for putting them on the market must be specified. Due to the seasonal restrictions of field seedling cultivation, supply is mainly concentrated in autumn and spring. Bulk supply in high temperature and cold seasons should be avoided to reduce the culture cost.

## (III) Considerations for Formulating the Production Plan

(1) Make a realistic estimate of the multiplication coefficient for various plants.

(2) Techniques, including explant induction, intermediate propagule proliferation, rooting, and acclimatization, should be available for the whole process of plant tissue culture.

(3) Master or get familiar with planting time and growth procedures of various tissue culture plantlets.

(4) Master the possible late-phase effects of tissue culture plantlets.

**Knowledge Points**

The proliferation rate of test-tube plantlets refers to the multiplication coefficient of intermediate propagules in the rapid propagation of plants. Estimate the propagation quantity of test-tube plantlets by taking plantlets, buds, or unrooted young shoots as the calculation unit (generally plantlets or culture flasks are used as the unit). The annual production ($Y$) depends on the number of plantlets per flask ($m$), the multiplication coefficient per cycle ($X$), and the number of annual proliferation cycles ($n$). It is expressed by the formula of $Y = mX^n$.

For instance, if propagation is carried out 8 times per year ($n = 8$), with a propagation coefficient of 4 each time ($X = 4$) and 8 plantlets per flask ($m = 8$), the annual propagable plantlets are: $Y = 8 \times 4^8 \approx 52 \times 10^4$. This is a theoretical value. In actual production, there are other factors, such as contamination and abnormal culture conditions, that will cause some losses. In addition, the equipment capacity and manpower are also restricting factors. For example, it is impossible to increase the number of flasks geometrically, nor is it possible to increase the number of employees responsible for inoculation and medium preparation in this manner. Therefore, the actual quantity produced should be lower than the estimated number.

## (IV) Production Planning

Although all plants have certain market demands, the demand varies in terms of timing and quantity, and explants may be collected in different seasons. Therefore, in order to ensure production and supply all year round, the overall annual sales goal and annual production plan must be decomposed into monthly sales plans and corresponding monthly production plans. The annual production plan should be broken down flexibly by fully considering the low and high seasons of the annual production of tissue culture plantlets. Priority should be given to the production plans for orders placed. In addition, attention should be paid to combining the varieties reasonably and preparing the production plan centering on users and the market. Once the production scale is determined, the culture cycle and the corresponding culture time are calculated according to the multiplication coefficient to arrange the specific production process. The production date should be determined with reference to the sales plan and sales period.

**Knowledge Points**

In principle, the release date of tissue culture plantlets should be 40–60 days earlier than the date of sales according to the plantlet varieties. The production quantity should be considered based on factors such as losses arising from contamination, elimination of variants and malformed plantlets, and survival rate by transplanting, and it should generally increase by 20%–30% higher than the planned sales volume. The formula for the calculation of the production quantity of tissue culture plantlets is as follows:

Annual (monthly) planned sales volume = Annual (monthly) actual production quantity × [1–Loss (generally 5%–10%)] × Survival rate by transplanting

For example, if the proliferation cycle for a certain tissue culture plantlet lasts 30 days on average, 1,200 plantlets/person/day can be transferred on a clean bench, and the ratio of proliferated to rooted plantlets is 3 : 7, with 5% of losses and a survival rate by transplanting of 85%. The average monthly production and annual production can be calculated based on 25 working days per month.

Average monthly production = 1,200 (plants) × 70% × 25 × (1–5%) × 85% ≈ 16,958 (plants)

Annual production = 16,958 (plants) × 12 = 203,496 (plants)

In the following text, the tissue culture plantlets for *Gerbera jamesonii* are taken as an example to explain how the annual sales and production plans are prepared (Table 8-1, Table 8-2). The peak and low seasons of *Gerbera jamesonii* production throughout the year can be clearly inferred from the tables, which is convenient for production arrangement. However, since the production plan is made based on the market demand with certain predictability, it may not be completely correct. In the production process, it is also necessary to make appropriate adjustments in a timely manner according to the changes in market demand, so as to better promote the timely production and effective sales of tissue culture plantlets.

**Table 8-1 Annual Sales Plan of Different Varieties of Tissue Culture Plantlets for *Gerbera jamesonii*** $10^4$ plants/year

| Variety | Month | | | | | | | | | | | | |
|---|---|---|---|---|---|---|---|---|---|---|---|---|---|
| | 1 | 2 | 3 | 4 | 5 | 6 | 7 | 8 | 9 | 10 | 11 | 12 | Total |
| Total | 3 | 3 | 13 | 20 | 20 | 10 | 2 | 2 | 10 | 12 | 3 | 2 | 100 |
| Variety 1 | 1.5 | 1.2 | 5 | 10 | 6 | 2 | 1 | 0.6 | 4 | 5 | 0.8 | 1 | 38.1 |
| Variety 2 | 1 | 1 | 3 | 5 | 5 | 5 | 0.5 | 0.3 | 3 | 3 | 1.2 | 0.3 | 28.3 |
| Variety 3 | 0.5 | 0.8 | 5 | 5 | 9 | 3 | 0.5 | 1.1 | 3 | 4 | 1 | 0.7 | 33.6 |

**Table 8-2 Annual Production Plan of Different Varieties of Tissue Culture Plantlets for *Gerbera jamesonii*** $10^4$ plants/year

| Variety | Month | | | | | | | | | | | | |
|---|---|---|---|---|---|---|---|---|---|---|---|---|---|
| | 1 | 2 | 3 | 4 | 5 | 6 | 7 | 8 | 9 | 10 | 11 | 12 | Total |
| Total | 3.6 | 3.6 | 15.6 | 24 | 24 | 12 | 2.4 | 2.4 | 12 | 14.4 | 3.6 | 2.4 | 120 |
| Variety 1 | 1.8 | 1.44 | 6 | 12 | 7.2 | 2.4 | 1.2 | 0.72 | 4.8 | 6 | 0.96 | 1.2 | 45.72 |
| Variety 2 | 1.2 | 1.2 | 3.6 | 6 | 6 | 6 | 0.6 | 0.36 | 3.6 | 3.6 | 1.44 | 0.36 | 33.96 |
| Variety 3 | 0.6 | 0.96 | 6 | 6 | 10.8 | 3.6 | 0.6 | 1.32 | 3.6 | 4.8 | 1.2 | 0.84 | 40.32 |

## II. Implementation of Production Plan

### (I) Establishment of Asexual Propagation System

The first step in implementing a production plan is to prepare the reproduction materials and achieve the desired proliferation base. First of all, it is necessary to ensure that the inoculated materials are from superior individual plants or lines with clear provenance and typical variety characteristics and without diseases, pests, or macroscopic viral symptoms. Any mistake made at the beginning of material collection will incur irreparable losses to the production. Secondly, after multiple reproductive buds are formed after primary induction culture, relevant varieties should be subject to virus detection and those carrying viruses should be eliminated. Special, rare, and precious varieties may not be used as reproduction materials before they become virus-free. Finally, the asexual propagation system is established through the experimental study in each culture stage. It can be said that the reproduction materials and tissue culture techniques lay a foundation for the production of commercial tissue culture plantlets.

### (II) Control of the Total Number of Proliferation Flasks on Shelf

When the qualified culture materials reach the required base after proliferation, the control of the total number of proliferation flasks on a shelf becomes crucial. The inventory should not be too much or too little. If proliferation is carried out blindly, subsequent work cannot be handled due to a lack of manpower or equipment after a certain period. The results will be a backlog of proliferation materials, some aged plantlets for optimal transfer and subculture, thin and poorly-grown plantlets for rooting, seriously reduced quality of final plantlets and the survival rate of transitional plantlets, weakened growth of reserved reproductive plantlets, and lowered proliferation rate, increasing the production cost and seriously affecting the quality of plantlets. On the contrary, an insufficient number of proliferation flasks on a shelf will cause insufficient reproductive mother plants, resulting in failure to accomplish the production plan on time, delaying the plantlet production period, and causing great economic losses.

**Knowledge Points**

The following factors should be comprehensively considered for the calculation of the total number of proliferation flasks on shelf: the number of production plans, the rooting ratio of each variety, and the work efficiency of operators. Their relationships are expressed as follows:

$$\text{Total number of proliferation flasks on shelf} = \frac{\text{Number of plantlets planned for production per month}}{\text{Number of plantlets produced in each proliferation flask per month}}$$

$$\text{Number of plantlets planned for production per month} = \text{Number of plantlets produced by each operator per day} \times \text{Working days per month} \times \text{Number of operators}$$

The number of plantlets produced in each proliferation flask per month refers to the number of plantlets for rooting within one month, which is related to factors such as the growth cycle and multiplication coefficient of the plantlets for tissue culture. For plantlets with a long growth cycle, the number of transfers within one month is small, i.e. the number of plantlets for rooting is small, and vice versa. If the proliferation rate and rooting rate are high, the number of mother plant flasks required per working day will be small, and the number of plantlets produced will be large. Conversely, if the proliferation rate is low, the best materials should be used for propagation to maintain the original number of proliferation flasks, resulting in fewer plantlets for rooting and fewer plantlets produced. As such factors are related to the production efficiency of tissue culture plantlets, it is extremely important to timely adjust the type and dosage of phytohormones in the medium and appropriately adjust the culture conditions according to the actual proliferation of a specific variety during production to improve the proliferation rate effectively.

According to the data calculated by the above formula, controlling the total number of proliferation flasks during the production of tissue culture plantlets helps realize the replacement of medium once for all reproductive plantlets in the proliferation stage within one cycle, so that the plantlets are kept in the optimal state in different growth stages, which is conducive to improving the quality of plantlets. According to the total number of proliferation flasks and the work efficiency of operators, the manpower required for production can be calculated and then properly arranged in the early stage of production to ensure smooth progress during production. For example, in an organization engaged in the mass production of tissue culture plantlets, several types of tissue culture plantlets will be produced in a year according to the market demand. Among them, the number of plantlets of *Gypsophila paniculata* to be produced in March is 120,000. According to the proliferation ability of tissue culture plantlets of *Gypsophila paniculata*, about 20 finished plantlets can be formed by rooting of each flask of proliferated plantlets in the subculture each time. If the proliferation cycle of tissue culture plantlets for *Gypsophila paniculata* is 15–20 days, and subculture can be carried out 1.5 times within one month, so it can be calculated that 4,000 flasks of proliferated plantlets should be prepared for production in March. The inoculation operation of *Gypsophila paniculata* is relatively simple. About 60 flasks of proliferated plantlets can be inoculated by one operator per day, and 1,000 flasks of proliferated plantlets can be inoculated within one proliferation cycle. Therefore, only four operators are required to complete the production of tissue culture plantlets for *Gypsophila paniculata*.

Since the products obtained through tissue culture and rapid propagation are viable plantlets, common problems such as poor growth, vitrification, chlorosis, and contamination of proliferated plantlets may occur in the production process, and the production plan may vary from month to month. Therefore, the data calculated as per the formula is only for reference. In actual production, the number of flasks of proliferated plantlets, together with the number of operators, should be adjusted timely according to specific circumstances.

## Task 2 Design of Industrialized Production of Tissue Culture Plantlets and Corresponding Technical Procedures

In the industrialized production of tissue culture plantlets, production procedures should be formulated and strictly implemented. Management and standard operation should be strengthened as required by the production plan to ensure the smooth connection of all technical procedures. Only in this way can high-quality tissue culture plantlets be timely and quantitatively produced to meet the market demand. There are five major technical procedures in the industrialized production of tissue culture plantlets: provenance selection, rapid propagation *in vitro*, acclimatization and transplanting, quality evaluation, packaging and transportation.

## I. Provenance Selection

Provenance is a prerequisite and the first consideration for the industrialized production of tissue culture plantlets. The selected plant varieties should not only meet the market demand but also local environmental conditions, so as to simplify production and reduce production costs.

**Knowledge Points**

Provenance can be acquired in two ways. The first is to obtain sterile original plantlets through outsourcing, technology transfer, or plantlet exchange. Purchased original plantlets are generally popular test-tube plantlets or those with small market potential and plantlet exchange is backed by strong technical strength. For tissue culture organizations with technical strength, technology transfer is preferred. Obtaining provenance in this way is convenient, time-saving, and shortens the rapid propagation process. This way is preferred when there is a large market demand that requires large-scale production within a short time. The second is through independent research and development, i.e. obtaining sterile original plantlets from the primary culture of explants. The two technical procedures to obtain sterile flask plantlets are selecting explants (generally apical and axillary buds) according to different culture purposes and plant species and disinfecting them accordingly. Building germplasm nurseries, strengthening variety breeding and mother plant breeding are the necessary measures to ensure pure provenance, convenient collection of inoculation materials, and timely update of the asexual reproduction system for tissue culture.

## II. Rapid Propagation In Vitro

It is about obtaining robust rooting plantlets through technical procedures and processes such as the primary culture for obtaining sterile materials, subculture for rapid propagation and proliferation, and rooting. This technical procedure also involves medium preparation, inoculation, and culture. The production process of tissue culture plantlets is shown in Fig. 8-1.

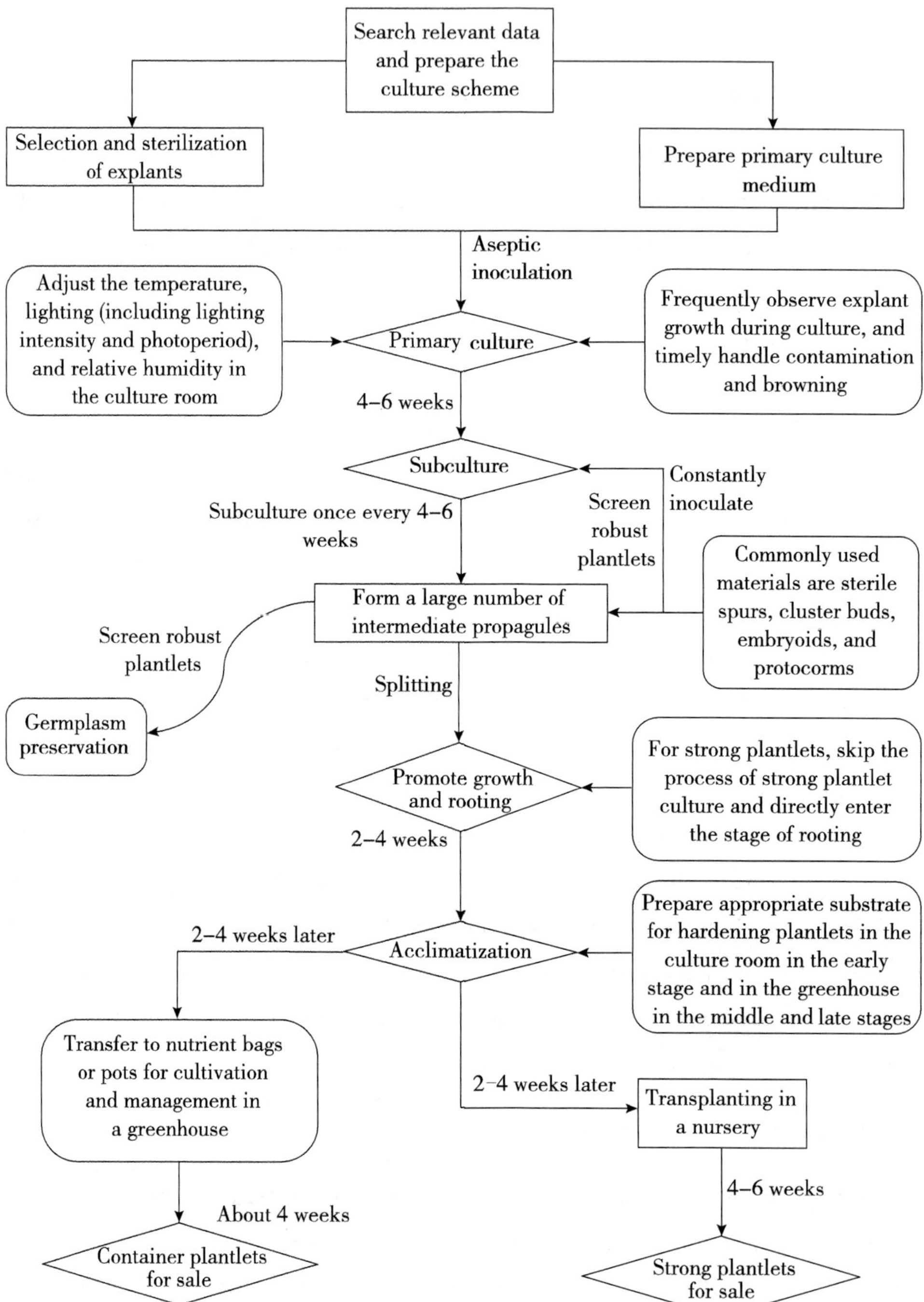


**Fig. 8-1 Production Process of Tissue Culture Plantlets**

## III. Acclimatization and Transplanting

The acclimatization of tissue culture plantlets aims to improve their adaptability to the natural environment with soilless cultivation technology before transplantation. This critical procedure determines the success of tissue culture and the ability to meet market demands in a timely manner. The procedures for acclimatization and transplanting of tissue culture plantlets are shown in Fig. 8-2.

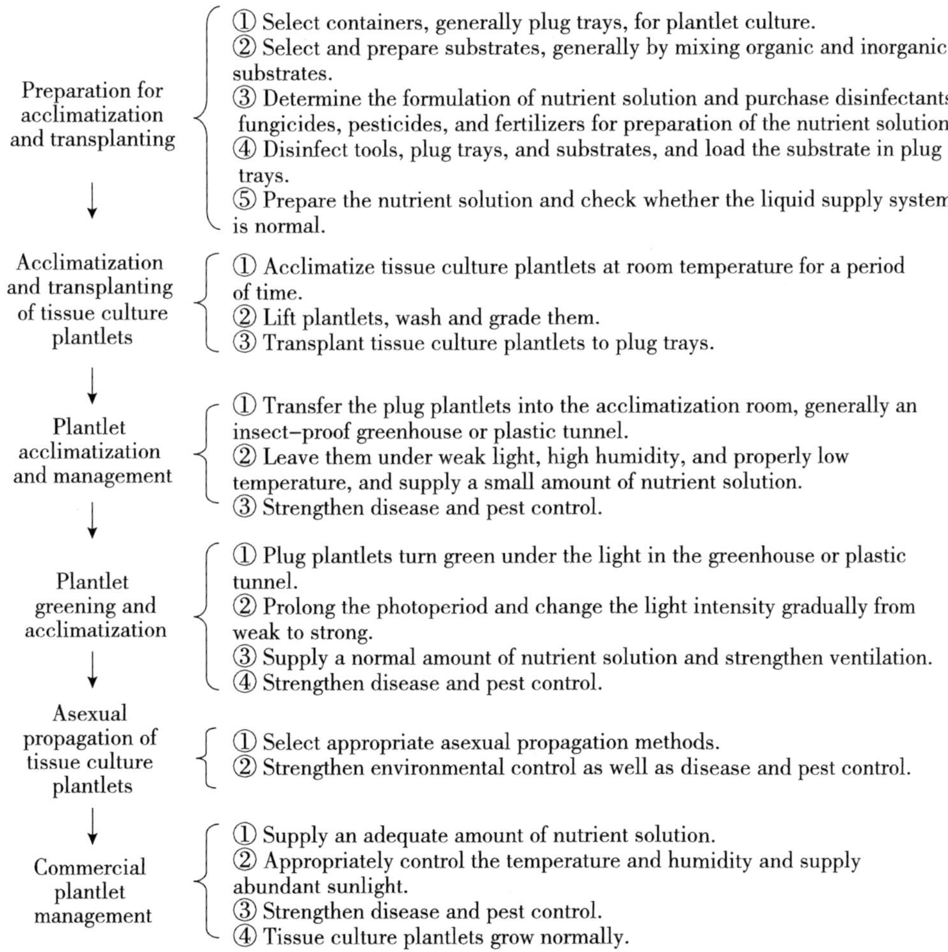


**Fig. 8-2 Operation Procedures for Acclimatization and Transplanting of Tissue Culture Plantlets**

## IV. Quality Evaluation

### (I) Items of Quality Evaluation of Tissue Culture Plantlets

Plantlet quality evaluation is an important link to ensure plantlet quality and safeguard the interests of planters. It is also an important basis for determining plantlet prices. With the promotion of tissue culture techniques, more and more tissue culture plantlets are cultivated and distributed as commodities. Considering the innovative production mode and the advanced products, quality inspection is required to be carried out strictly. Currently, quality inspection criteria for tissue culture plantlets in China are not yet perfect, and many companies and institutions specializing in the quality inspection of tissue culture plantlets have emerged in the United States.

**Knowledge Points**

Quality evaluation of tissue culture plantlets mainly involves the following aspects.

1. Commodity Traits

(1) Plantlets with better prematurity and quality will be rated with a higher grade.

(2) Agronomic traits, such as leaves, growth, plant height, stem diameter, plant expansion and root system, should be graded according to specific requirements of different crops.

2. Health Status

(1) Whether the plantlets carry the epidemic pathogens, fungi and bacteria.

(2) Whether they carry viruses.

3. Genetic Stability

(1) Whether there are typical traits of the variety.

(2) Whether they are neat and consistent.

(3) Rapid propagation materials are identified through DNA "fingerprinting" techniques, such as RAPD or AFLP, to determine their genetic stability.

## (II) Quality Criteria for Tissue Culture Plantlets

The quality criteria for tissue culture plantlets as provenance are to be free of viruses or pathogens and pure in terms of varieties. The quality criteria are based on four indicators: the root system, general condition, plantlet height, and the number of leaves. The sequence of the importance of each index is root system status > general condition > height of plantlets for transplanting > the number of leaves. Rootless plantlets with poor growth and in black color should be rejected and evaluated as unqualified. Only when the plantlets are qualified in terms of the root system can they be further evaluated according to other indexes. See Table 8-3 for the quality criteria for tissue culture plantlets of several common flowers.

**Table 8-3 Quality Criteria for Tissue Culture Plantlets of Several Common Flowers**

| Name | Grade | Root System Status | General condition | Plantlet Height (cm) | Number of Leaves |
|---|---|---|---|---|---|
| *Gypsophila paniculata* | Grade 1<br>Grade 2 | Rooted<br>With root primordium or rootless | Plantlets are stout and straight, with dark green leaves | 2–3<br>1.5–3 | 4–8<br>4–8 |
| *Gerbera jamesonii* | Grade 1<br>Grade 2 | Rooted<br>Rooted | Plantlets are upright and solitary, with green leaves; The central buds are smaller than those of Grade 1 plantlets. For some plantlets, the leaves are irregular and there are central buds | 2–4<br>1–3 | >3<br>>3 |
| *Myosotis sylvatica* | Grade 1<br>Grade 2 | Rooted<br>Rooted | Plantlets are solitary, with green leaves and central buds | 2–3<br>2–4 | >3<br>>3 |
| *Eustoma ressellianum* | Grade 1<br>Grade 2 | Rooted<br>Rooted | Plantlets are solitary, with green leaves, and without rosettes | 2–4<br>1.5–3 | >6<br>4–6 |

continued

| Name | Grade | Root System Status | General condition | Plantlet Height (cm) | Number of Leaves |
|---|---|---|---|---|---|
| Chrysanthemum | Grade 1<br>Grade 2 | Rooted<br>Rooted | Plantlets are stout and straight, with grayish-green leaves | 2–4<br>1–2 | >4<br>>4 |
| French marigold | Grade 1<br>Grade 2 | Rooted<br>Rooted | Plantlets are stout and upright, with green leaves | 3–4<br>1–3 | >5<br>>3 |

Tissue culture plantlets should be transplanted after they survive acclimatization. The quality of plantlets for transplanting will affect the survival rate, growth, and yield, as well as disease and pest control. The quality criteria for transplanting plantlets are mainly based on the stalk diameter, plantlet height, root system status, number of leaves, general condition, and uniformity, as well as disease and pest damage. See Table 8-4 for the common quality criteria for plantlets for transplanting, and Table 8-5 for the quality criteria for several common cut flowers for transplanting.

**Table 8-4 Common Quality Criteria for Plantlets for Transplanting**

| Evaluation Item | | Grade | | |
|---|---|---|---|---|
| | | Grade 1 | Grade 2 | Grade 3 |
| 1 | Root system status | Roots grow uniformly and completely, without defects | Roots grow relatively uniformly and completely, with no or minor defects | Roots are intact and grow moderately, with minor defects |
| 2 | General condition | Plantlets grow vigorously. They are intact and uniform in shape, and look fresh, stout, straight, and symmetrical, with greenish and glossy leaves | Plantlets grow normally. They are intact and uniform in shape, and look fresh, relatively stout, straight, and symmetrical, with greenish and slightly glossy leaves | Plantlets grow moderately. They are intact and uniform in shape, and look slightly fresh, with green and slightly glossy leaves. They may overgrow to a greater or lesser extent |

continued

| Evaluation Item | | Grade | | |
|---|---|---|---|---|
| | | Grade 1 | Grade 2 | Grade 3 |
| 3 | Uniformity | For more than 90% of plantlets of the same grade, their ground diameter and height are within ± 10% of the average ground diameter and the average height of the same batch of plantlets | For more than 85% of plantlets of the same grade, their ground diameter and height are within ± 10% of the average ground diameter and the average height of the same batch of plantlets | For more than 80% of plantlets of the same grade, their ground diameter and height are within ± 10% of the average ground diameter and the average height of the same batch of plantlets |
| 4 | Disease and pest damage | No quarantine diseases or pests, and no spots caused by diseases and pests | No quarantine diseases or pests, and no spots caused by diseases and pests | No quarantine diseases or pests, and no spots caused by diseases and pests |

## V. Packaging and Transportation

### (I) Requirements for Packing Boxes

The quality of packing boxes may vary according to the type of plantlets and transportation distance. For short-distance transportation, simple carton boxes or wooden crates can be used to reduce packaging costs; For long-distance transportation, plantlets should be placed in multiple layers to make full use of the space, and the capacity and strength of the boxes should be considered to ensure that they can withstand pressure and bumpiness. The packing boxes should be printed with a company's trademark to facilitate the sales of tissue culture plantlets by the company.

### (II) Requirements for Tissue Culture Plantlets

As tissue culture plantlets are generally subject to hydroponics and substrate culture after surviving acclimatization, all their roots will be exposed once they are lifted. Therefore, roots must be kept moisturized to avoid the

**Table 8-5 Common Quality Criteria for Several Common Cut Flowers for Transplanting**

| S/N | Variety | Grade 1 | | | Grade 2 | | | Grade 3 | | |
|---|---|---|---|---|---|---|---|---|---|---|
| | | Ground Diameter (cm) | Plantlet Height (cm) | Number of Leaves (piece) | Ground Diameter (cm) | Plantlet Height (cm) | Number of Leaves (piece) | Ground Diameter (cm) | Plantlet Height (cm) | Number of Leaves (piece) |
| 1 | *Gypsophila paniculata* | ⩾0.6 | 6–8 | ≥14 | 0.4–0.6 | 5–6 | 11–12 | 0.2–0.4 | 4–5 | 10–11 |
| 2 | *Gerbera jamesonii* | ⩾0.5 | 10–12 | ≥10 | 0.4–0.5 | 8–10 | 4–7 | 0.3–0.4 | 6–8 | 3–4 |
| 3 | *Limonium sinense* | ⩾0.5 | 12–14 | ≥14 | 0.3–0.5 | 10–12 | 8–10 | 0.2–0.3 | 8–10 | 6–8 |
| 4 | *Desmodium gyrans* | ⩾0.5 | 11–13 | ≥12 | 0.3–0.5 | 9–11 | 6–12 | 0.2–0.3 | 6–9 | 4–6 |
| 5 | *Eustoma ressellianum* | ⩾0.4 | 6–8 | ≥10 | 0.3–0.4 | 4–6 | 6–8 | 0.2–0.3 | 3–4 | 4–6 |
| 6 | Chrysanthemum | ⩾0.6 | 8–12 | ≥12 | 0.5–0.6 | 6–8 | 10–12 | 0.4–0.5 | 3–4 | 8–10 |
| 7 | French marigold | ⩾0.5 | 8–12 | ≥14 | 0.4–0.5 | 5–6 | 10–14 | 0.2–0.4 | 3–5 | 6–10 |

impact on survival rate after long-distance transportation. Rock wool and peat should be used as substrates as they are lightweight and have moisturizing effects, which are conducive to effective root protection. The currently promoted plug tray cultivation of plantlets requires a small amount of substrate and can effectively protect roots. In addition, plug trays are easy to pack, making them suitable for plantlet transportation. Generally, plantlets, especially balled ones, are appropriate for long-distance transportation. Plantlets at a young age with a few leaves are not easily damaged during transportation, and the transportation cost per plantlet is low. However, in the early stage, when production significantly affects the production value, the plantlets cultivated for early maturity for protected fields and open fields in spring should be old enough to meet the requirements of users.

### (III) Requirements for Transportation Means

The means of transport should be selected based on distance. Generally, trucks are applicable for short-distance transportation, while trains or large-capacity trucks are for long-distance transportation, preferably with temperature and humidity control devices. It is best to transport the tissue culture plantlets directly to the planting sites and avoid repeated loading and unloading to reduce plantlet damage. Precious plantlets or those in urgent need can also be transported by air.

### (IV) Requirements for Transportation Temperature

Generally, the tissue culture plantlets should be transported at low temperatures (9–18°C). Fruit and vegetable plantlets should be transported at an optimal temperature of 10–21°C (lower than 4°C or higher than 25°C are not suitable). Cold-resistant leaf vegetable plantlets such as head cabbage should be transported at 5–6°C.

### (V) About Transportation

1. Preparation for Transportation

(1) The specific departure date should be set, and nurseries and users

should be timely notified to pay attention to weather forecasts. Good preparation should be made before transportation, especially in winter and spring, to protect plantlets against cold and frost. Plantlets should be acclimatized a few days before transportation by gradually lowering the temperature, and appropriately supplying less or no nutrient solution to enhance their stress resistance.

(2) Packaging should be accelerated to minimize the time, reduce repeated loading and unloading of plantlets, and minimize plantlet damage.

(3) To ensure and improve the survival rate of transported tissue culture plantlets, attention should be paid to root protection and root treatment. Hydroponic plantlets or substrate-cultured plantlets generally have no substrate attached to them once lifted. They can be tied into a bundle of dozens to hundreds of plantlets (depending on the plantlet size), with roots wrapped with sphagnum or other moisturizing packaging materials, and then packed. The plug plantlets are transported with substrates. They may be placed directly in boxes after being pulled out of the plug trays by shaking with intact roots or they may also be dipped in the mud mixed with nutrient solution for root protection, and then wrapped with plastic film for moisturization, so as to improve the survival and recovery rate after field planting.

2. Notes for Transportation

Fast and punctual transportation is required and long-time lingering is not recommended during long-distance transportation. Plantlets should be handed over to users as soon as possible upon arrival to allow them to be planted timely. If plantlets are transported by a vehicle with temperature and humidity control measures, attention should be paid to controlling the temperature and humidity to prevent the plantlets from being damaged when the temperature and humidity get too high or too low.

## Task 3 Cost Accounting for Industrialized Production of Tissue Culture Plantlets and Production Management

### I. Cost Accounting

The industrialized production of tissue culture plantlets is for commercial purposes. Only by producing test-tube plantlets with super quality and competitive price can a company win in the market competition and obtain good economic benefits. Therefore, the production costs must be accounted for. The cost index of a factory for the production of tissue culture plantlets is an overall index reflecting its management level and work quality. It is not only the basis for understanding various consumptions in production, improving technological processes, and improving weak links, but also a necessary measure to improve benefits and save investment. The cost accounting for the production of tissue culture plantlets is relatively complex as it features characteristics of both industrial production (annual indoor planting) and agricultural production (greenhouse or field planting). Under the influence of climate and season, long-time management is needed before tissue culture plantlets are ready for sale. In addition, the propagation coefficient and growth rate of different species and varieties vary greatly, making it difficult to carry out accurate cost accounting item by item. The general practice is to carefully record the expenditures of a certain number of tissue culture plantlets in annual production. See Table 8-6 for the cost breakdown of commercialized production of tissue culture plantlets. See Table 8-7 for the cost accounting of production workshop with an annual yield of one million test-tube plantlets.

**Table 8-6 Cost Breakdown of Commercialized Production of Tissue Culture Plantlets**

<table>
<tr><th>Name</th><th colspan="2">Specific Item</th><th>Remarks</th></tr>
<tr><td rowspan="2">Sales Costs</td><td>Direct Costs</td><td>Costs for preparation of media, water and electricity charges, payroll, etc.</td><td>The costs directly allocated for the production of tissue culture</td></tr>
<tr><td>Indirect Costs</td><td>Depreciation and maintenance costs of instruments and equipment, production material costs, etc.</td><td>Plantlets can be included in a proportion in the production cost of tissue culture plantlets</td></tr>
<tr><td colspan="2">Indirect Costs</td><td>Management costs, sales costs, financial costs, etc.</td><td>The costs incurred from the organization and management of production must not be included in the production cost of tissue culture plantlets</td></tr>
</table>

Table 8-7 **Cost Accounting of Production Workshop with an Annual Yield of One Million Test-tube Plantlets** CNY

| Production Process | Capital Costs | Equipment Depreciation Costs | Reagent Costs | Water Charges | Electricity Charges (Heating Fee) | Payroll | Total |
|---|---|---|---|---|---|---|---|
| Washing | 7,150 | 16,700 | — | 2,000 | 200 | 4,500 | 30,550 |
| Medium Preparation | — | — | 2,306 | — | 10,608 | 6,918 | 19,832 |
| Inoculation | — | — | — | — | 1,844 | 28,825 | 30,669 |
| Culture | — | — | — | — | 31,738 | — | 31,738 |
| Total | 7,150 | 16,700 | 2,306 | 2,000 | 44,390 | 40,243 | 112,789 |
| Proportion of total expenses(%) | 6.34 | 14.81 | 2.04 | 1.77 | 39.36 | 35.68 | — |

## II. Measures for Improving Economic Benefits of Tissue Culture Plantlets

Whether the industrialized production of tissue culture plantlets can achieve good economic benefits mainly depends on market factors and management level. Famous, special, new, and good-quality tissue culture plantlets should be produced in time according to sales and market demand and put on the market in batches, so as to reduce costs and effectively improve economic benefits. In addition, strengthening production management, reducing production costs, or adding value to tissue culture plantlets are important measures to improve the economic benefits.

Improving economic benefits mainly considers two aspects: reducing production costs and improving proliferation efficiency.

### (I) Reducing Production Costs

The production costs of tissue culture plantlets are affected by many factors, but it mainly depends on equipment conditions, management level, and the proficiency of operators.

(1) Reduce the pollution rate, appropriately increase the propagation coefficient and subculture generations, and improve the rooting rate and the survival rate of transplanted plantlets.

(2) Improve the operation level and production efficiency of inoculators to reduce labor costs.

(3) Reduce equipment investment, correctly use instruments and equipment, prolong service life and reduce maintenance costs.

(4) Minimize energy consumption and make full use of the culture room.

(5) Simplify tissue culture procedure, save production materials, facilitate operation and improve the survival rate of transplanted plantlets.

(6) Carry out moderate-scale production to improve profits.

(7) Strengthen collaboration with peers to diversify the types of tissue culture plantlets.

### (II) Improving Proliferation Efficiency

(1) Cultivate rare, famous, special, good-quality, and virus-free plantlets to capture the market.

(2) Cultivate tissue culture plantlets with independent property rights and adopt brand management strategies.

(3) Sell sieve tray plantlets or nutrition pot plantlets.

(4) Carry out the division and cutting propagations with tissue culture plantlets that survive acclimatization.

(5) Carry out both the sales and medicinal ingredient extraction of plantlets.

## III. Operation and Management Thoughts and Measures

### (I) Operation and Management Thoughts

Operation and management thoughts of a company run through the whole process of its operation and guide the company to adjust external and internal activities and formulate policies and objectives according to market needs and changes, thus helping the company to survive and thrive. Under the guidance of operation and management thoughts, the operation and management theory is formed to guide the production and management practice and continuously promote the productivity improvement of enterprises.

The industrialized production of tissue culture plantlets combines cultivation, production, and sales of tissue culture plantlets, and is subject to agriculture-industry-commerce integrated operation and enterprise-based management. Against the backdrop of a market economy, in terms of the commercialized production of tissue culture plantlets, market awareness and market-based operation thoughts should be cultivated. Production should be planned according to the market demand and sales. Technological innovation should be strengthened to improve the quality of tissue culture plantlets and to enhance enterprises' market competitiveness. People-oriented, scientific and standard management should be advocated by establishing sound internal management systems and mechanisms and by emphasizing cost accounting. High-quality plantlets should be produced on a large scale to maximize economic benefits. Customers should be placed in the first place and market survey should be conducted. After-sales services for tissue culture plantlets should be provided to keep track of the changes in market demand and arrange production properly, so as to continuously meet the demands of distributors and customers, and remain invincible on the market.

## (II) Operation and Management Measures

Operation and management measures for the industrialized production of tissue culture plantlets include institutional setting, formulation and implementation of production plans, personnel management, production process management, production management, and sales management.

1. Institutional Setting and Job Responsibilities

For the industrialized production of tissue culture plantlets, proper production plans should be formulated and effectively implemented, talent and technology advantages should be put into full play, and production efficiency should be continuously improved. However, all these are directly affected by the institutional setting, management system, and regulations. Therefore, it is necessary to build a proper organizational structure and sound management system, and clarify job responsibilities for the industrialized production of tissue culture plantlets. The organization structure setting and the job responsibilities of the main departments of the tissue culture factory are shown in Fig. 8-3 and Table 8-8, respectively.

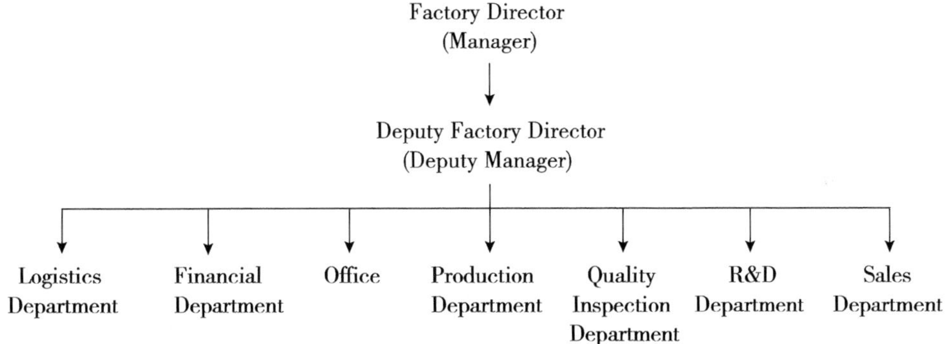


**Fig. 8-3 Organizational Structure Setting of Tissue Culture Factory**

2. Preparation and Implementation of Production Plans

The production plan should be scientifically formulated and effectively implemented based on the factors of market survey, analysis, and predictions, in combination with the production conditions and capacity of the workshop so as to achieve reasonable, targeted, and efficient production.

3. Personnel Management

Personnel management is an important part of industrialized production management of tissue culture plantlets. Good personnel management can ensure production efficiency and product quality and improve the economic benefits. In terms of management thoughts, management should be people-oriented and all employees can do their best. In terms of employee management, technical training of employees should be strengthened to enable them to meet the requirements of skilled, rapid, and accurate operation with a low contamination rate. The production responsibility system should be implemented among employees, making each of them responsible for a certain production procedure and achieving the corresponding quota. The salary of employees should be linked to corporate revenues, calculated based on piece rate, and raised or deducted according to their performance. These are effective measures to improve labor

**Table 8-8 Job Responsibilities of Main Departments of Tissue Culture Factory**

| Department Name | Type of Work | Job Responsibilities |
|---|---|---|
| Production Department | Handypersons and cleaners | 1. Wash utensils and tools.<br>2. Keep public sanitation in production areas.<br>3. Pack tissue culture plantlets, load and unload raw materials, and finished plantlets |
| | Medium preparation workers | 1. Prepare media for production.<br>2. Safeguard drugs and maintain instruments and equipment |
| | Inoculation workers | 1. Cut, transfer and inoculate culture materials.<br>2. Keep the buffer room and inoculation workshop clean.<br>3. Sterilize the inoculation workshop regularly |
| | Culture workers | 1. Set and control culture conditions in the culture workshop.<br>2. Keep daily records about warehouse-in and warehouse-out of culture materials.<br>3. Timely remove contaminated and abnormally grown and differentiated plantlets (flasks).<br>4. Keep the culture workshop clean and disinfect it regularly |
| | Workers for acclimatization and transplanting | 1. Acclimatize tissue culture plantlets for transplanting.<br>2. Carefully transplant and manage plantlets, and keep records.<br>3. Keep records about warehouse-in and warehouse-out of plantlets.<br>4. Carry out daily management of acclimatizing and transplanting greenhouses, and keep them safe and sanitary |
| Quality Inspection Department | Quality inspectors | 1. Establish the workshop quality standard for tissue culture plantlets.<br>2. Inspect the quality of tissue culture plantlets |
| R&D Department | Developers | 1. Establish the asexual propagation system for tissue culture.<br>2. Study optimal medium formulation and culture conditions.<br>3. Resolve problems in production.<br>4. Finish relevant work concerning the provenance and make relevant techniques available.<br>5. Establish technical files and keep them confidential |
| Sales Department | Sellers | 1. Carry out relevant work concerning publicity and advertising.<br>2. Conduct market surveys and make market predictions.<br>3. Carry out sales negotiation and reach sales targets.<br>4. Provide excellent after-sales services for tissue culture plantlets |
| Logistics Department | Logistics workers | 1. Purchase instruments, equipment, and production supplies.<br>2. Ensure the normal supply of water and electricity.<br>3. Repair and maintain instruments and equipment |

productivity. They can not only improve the work efficiency of employees but also motivate them in production and innovation, contributing to the growth of enterprises. Managers should have a strong sense of responsibility and be qualified and capable in management. Their work efficiency can be measured according to ISO 9000 standards by organizing competitions within the enterprise.

4. Production Process Management

The industrialized production processes of tissue culture plantlets are relatively complex and involve a lot of work. By formulating reasonable rules and regulations and implementing scientific and standardized management, production can be carried out as planned, personal and property losses caused by human errors can be avoided, and product quality can be ensured. The main rules and regulations are formulated as follows: explant collection and propagation registration system; technical procedure for aseptic operation; technical procedure for medium preparation; operating procedure for autoclaves; management system for acclimatization and transplanting of tissue culture plantlets; inspection and quarantine system for mother plants and commercial plantlets; employment management system; reward and punishment system; post responsibility system and production quota management system. In addition, each department should prepare the "Operation Instructions" and the staff should complete the work tasks in strict accordance with the "Operation Instructions". The "Product Release Criteria" should be established for handover between different departments to avoid the release of unqualified finished or semi-finished products to the next production procedure, so as to ensure that the released products meet the specified standards. There should be written records and responsible persons for tracking upon completion of each process. The whole production process can be managed by computers.

5. Product Management

A complete archive should be established for each kind of tissue culture plantlets, indicating the traits and planting (sampling) place of mother plants, inoculation date, subculture generations, production quantity, sales place, and growth status. Each product should be uniquely numbered to facilitate inquiry and identification and differentiation in the production process, so as to ensure product quality and after-sales tracking service.

6. Sales Management

Operators should conduct market surveys, analysis, and prediction; determine product publicity and promotion strategies; understand users' demands, keep orders well and complete them; determine a reasonable scope of sales, and select proper sales channels and sales modes (Table 8-9) according to such factors as corporate conditions, and the type, quantity, supply and demand relationship, and price of products; sell products as many as possible, recover funds as soon as possible to reduce business risks; provide convenience for customers as much as possible based on the principle of users first; carry out product sales statistics, after-sales services, and tracking surveys.

**Table 8-9 Basis for Selection of Sales Mode for Tissue Culture Plantlets**

| Main Sales Mode | Basis for Selection |
|---|---|
| Personal Selling | 1. Centralized plantlet market.<br>2. Small companies, low production, insufficient funds.<br>3. Centralized supply of local plantlets, short transportation distance, and good sales timeliness |
| Advertising | 1. Decentralized plantlet market.<br>2. Large companies, high production, and abundant funds.<br>3. Rare varieties and a small number of plantlets are used.<br>4. Within the trial marketing period of tissue culture plantlets.<br>5. Exhibition site |

7. Technical Exchange and Promotion

Enterprises should organize and arrange for technicians to participate in various academic conferences and further study; cooperate with universities, scientific research institutions, and enterprise peers, or organize social activities, so as to strengthen technical learning and exchange; timely learn advanced technologies, and draw on valuable experience to update and improve their technologies. In addition, by virtue of their advantages in talent and technology, enterprises should actively carry out technical training and technology promotion to improve social visiblity and influence, and obtain certain economic benefits while creating conditions for future projects and technical cooperation. Furthermore, this is also good for product sales. Operators should formulate and implement plans for technical exchange and promotion in a targeted, organized, and effective manner, making sure that a summary is made.

# Summary

1. Industrialized Production Techniques for Tissue Culture Plantlets

(1) Requirements for site selection: The factory should be located far away from pollution sources and in a place with convenient transportation, an uninterrupted supply of water and electricity, and easy drainage.

(2) Industrialized production techniques: Techniques for provenance selection, rapid propagation *in vitro*, acclimatization and transplanting, plantlet quality evaluation, packaging, and transportation.

2. Production Planning and Implementation

(1) Production planning: Breeding varieties, planned quantity, and roll-out time.

(2) Implementation of production plan: Keep a proper number of proliferation flasks on shelf and a proper ratio of proliferated to rooted plantlets, and correctly estimate the actual annual production.

3. Production Cost Accounting and Measures to Improve Benefits

(1) Production cost accounting: Labor cost, equipment depreciation cost, raw material cost, and water and electricity charges, among others.

(2) Measures to improve benefits: Improve labor productivity; reduce equipment investment and prolong service life; cut water and electricity expenses; reduce the use of flasks and use inexpensive substitutes; reduce contamination rate; improve propagation coefficient and survival rate by transplanting; diversify businesses and establish cooperations with relevant organizations in the industry.

4. Production Management and Operation

(1) Production management: Implement the production management responsibility system during production, and make sure that plantlets are produced and planted in their growing seasons.

(2) Operating strategy: Correctly predict the market; produce plantlets based on sales; integrate production and sale; build a corporate reputation to win benefits.

# Review Test

## I. Explain the Glossary

1. Market Share
2. Multiplication Coefficient of Test-tube Plantlets

## II. Fill in the Blanks

1. The rapid propagation procedures for plants are generally included in four stages: ________, _______, _______, and_______.

2. The production workshops for tissue culture plantlets generally include ________, _________, ________, ________, ________ , ________ , and testing workshop.

## III. True or False Questions

(  ) 1. During the rapid propagation of plants, the longer the culture cycle, the higher the multiplication coefficient each time, and the higher the annual multiplication coefficient.

(  ) 2. The proportion of plantlets in the flasks for proliferation and rooting depends on the multiplication coefficient. A higher proliferation rate means a larger proportion of rooting, fewer flasks of mother plants required per working day, and more plantlets produced.

(  ) 3. In the process of rapid propagation, if the propagation coefficient, rooting rate, and survival rate by transplanting are high, and the proliferation and rooting cycles are short, the production cost is low.

## IV. Multiple Choice Questions

1. The main objective at the stage of ______ is to maintain stable proliferation until reaching the quantity required.

A. Initiation of the establishment of a stable sterile culture system

B. Bud proliferation

C. Bud emergence and growth

D. Acclimatization and transplanting of test-tube plantlets

2. The theoretical annual proliferation of tissue culture plantlets can be calculated by the following formula: $Y=mX^n$, where $X$ represents:

A. Annual proliferative generations B. Initial number of explants

C. Multiplication coefficient per cycle D. Days of growth per generation

3. Four sterile plantlets are obtained from one plant by primary culture, and then subcultured once every 36 days, with a multiplication coefficient of 3 each time. Hence there will be ______ propagated plantlets after one year.

A. $3\times4^{10}$ plants B. $4\times3^{12}$ plants C. $4\times3^{10}$ plants D. $3\times4^{12}$ plants

4. Assume there are 12 flasks of test-tube plantlets, each containing 6 plantlets, and that the multiplication coefficient is 3 per cycle, totaling 8 multiplications a year. Theoretically, the annual total number of plantlets that can be propagated is about ________.

A. 472,400 B. 172,800 C. 47,240 D. 1,728

5. In the expanded propagation of a plant, 5 persons are responsible for inoculation every day, and each person inoculates 100 flasks every day. If one month is a proliferation cycle and it includes 20 working days, the total number of proliferation flasks on shelf is ________.

A.12,000 B.120,000 C.1,000 D.10,000

## V. Short Answer Questions

1. How to design the production workshop for tissue culture plantlets?

2. What are the criteria for quality evaluation of tissue culture plantlets?

3. How to analyze the economic benefits of tissue culture workshops?

4. In the production of test-tube plantlets, which main measures can reduce costs and improve economic benefits?

5. Briefly describe the management measures for plant tissue culture.

# Module 9 Application of Tissue Culture Technique in Transgenic Engineering

## Main Content

With limited resources, crop yield cannot be increased mainly by expanding cultivated area and irrigation area but should be increased by improving the yield per unit of crops. Genetic engineering can accurately transform plants, and significantly improve yield, resistance, and quality of transgenic crops. At the same time, it can also greatly reduce agricultural production costs and alleviate the deterioration of the agricultural ecological environment. Although the safety of transgenic crops has aroused fierce debate worldwide, with the continuous deepening of research in plant genetic engineering and the continuous improvement of technology, this problem will be gradually clarified and solved, contributing to agricultural production and development.

Receptors for plant genetic transformation are commonly called explants, which must be easily regenerated with high regeneration frequency and good stability. The establishment of an efficient and stable regeneration system is the key to the success of plant genetic transformation. According to the totipotency theory in plant tissue culture, somatic cells of any part of a plant can grow into a complete plant, but not all somatic cells can regenerate into complete plants in the practice of plant tissue culture. Some plants have not yet formed their own efficient and stable regeneration system. Therefore, tissue culture techniques, which are the basis of transgenic technology, should be used to build an efficient and stable regeneration system.

# Work Tasks

## Task 1 Overview of Plant Genetic Transformation Receptor System

Selecting and establishing a good plant receptor system is one of the critical factors influencing the success of the genetic transformation. The plant receptor system refers to a regeneration system through which the explants for transformation can efficiently and stably regenerate clones by tissue culture, can be integrated with exogenous genes, and are sensitive to the transformation of selective antibiotics.

### I. Conditions of Plant Genetic Transformation Receptor System

1. Efficient and Stable Regeneration Capacity

Explants for plant genetic transformation must be easily regenerated with high regeneration frequency, good stability, and good repeatability. Theoretically, somatic cells in any part of a plant are totipotent and can regenerate into plants. However, in practice, not all somatic cells can regenerate into complete plants. The genetic transformation frequency is generally low, only 0.1%. In addition, some treatments in plant genetic transformation such as the infection of *Agrobacterium tumefaciens*, transformant screening by antibiotics, and subculture, will reduce the regeneration frequency of transformed explants compared with non-transformed explants to varying degrees. To obtain a high transformation rate, the receptor system for genetic transformation should have a stable regeneration rate higher than 80%, and cluster buds must be regenerated on each explant (the more cluster buds, the better), making it possible to achieve high transformation efficiency.

2. High Genetic Stability

Plant genetic transformation is to purposefully introduce specific genes into plants and integrate, express, and inherit them to modify inferior traits. This requires that the plant receptor system should not be affected in terms of division and differentiation after receiving specific genes, and can stably pass exogenous genes to progeny, maintaining genetic stability and minimizing variation. Variations are common in plant tissue culture. They are related to the tissue culture methods, regeneration methods, and the type of explants. Therefore, these factors should be fully considered in the establishment of the genetic transformation receptor system, so as to ensure the genetic stability of transgenic plants.

3. Stable Source of Explants

To establish an efficient and stable regeneration system for genetic transformation, it is necessary to have a stable source of explants, that is, explants should be relatively easy to obtain in large quantities. Considering the low frequency of genetic transformation, tests should be carried out repeatedly. Thus, a large number of explant materials are required. Leaves, petioles, hypocotyls, roots of sterile tissue culture plantlets, and calluses produced from these organs are generally used as explants for transformation.

4. Sensitive to Selective Antibiotics

Antibiotic selective marker genes are artificially added during vector construction so that the genetically transformed cells or plants have resistance to some antibiotics and can grow, develop, and differentiate into regenerated plants on the selective medium with a certain concentration of antibiotics, while the non-transformed cells or plants cannot survive. The antibiotics used to screen transformants in genetic transformation are called selective antibiotics, which require plant receptor materials to be sensitive to them, i.e. when the selective antibiotics added to the selective medium reach a certain concentration, they can inhibit the growth, development, and differentiation of non-transformed plant cells. The transformed plant cells can grow, divide, and differentiate normally into completely transformed plants because they carry the resistance gene of the antibiotics.

5. Sensitive to *Agrobacterium* Infection

Currently, *Agrobacterium*-mediated transformation is the most commonly used method in plant genetic transformation. This method requires plant receptor materials to be sensitive to *Agrobacterium* so they can accept exogenous genes. Some monocotyledons and gymnosperms are not sensitive to *Agrobacterium* infection. Even dicotyledons vary in their degree of sensitivity to *Agrobacterium*, and different plants, even different tissue cells of the same plant, vary greatly in their sensitivity to *Agrobacterium* infection. Therefore, the sensitivity of the receptor system to *Agrobacterium* infection must be tested before the *Agrobacterium*-mediated transformation system is selected, and only plant materials sensitive to *Agrobacterium* infection can be used as the receptor system.

## II. Type of Plant Genetic Transformation Receptor System

Currently, various effective receptor systems have been established for use in different transformation methods and to meet different transformation purposes. During the specific operation in genetic transformation, they should be used based on plant species, genetic vector systems, experimental equipment, and other factors.

1. Protoplast Receptor System

Protoplast is the part of a plant cell with cytoderm removed. It has the ability to differentiate, multiply, and regenerate into intact plants under appropriate conditions and is totipotent. Since there is only a thin layer of cytomembrane that separates the protoplasm from its exterior environment, plant genetic transformation can be realized by changing the permeability of the cytomembrane with some physical or chemical methods, to allow exogenous DNA to enter the cell and integrate into the chromosome for expression. Therefore, the protoplast regeneration system is an ideal receptor system for genetic transformation.

2. Callus Regeneration System

The callus regeneration system refers to the receptor system in which the explant is dedifferentiated to produce calluses that then regenerate into plants. A callus is composed of the dedifferentiated meristem that easily receives exogenous DNA with a high transformation rate. Multiple explants can be induced by tissue culture to produce calluses that can be applied to the genetic transformation of multiple plants. Calluses can be subcultured for expanded propagation, so more transformed plants can be obtained by callus culture.

3. Directly Differentiated Bud Receptor System

Directly differentiated buds refer to the adventitious buds formed through direct differentiation of explant cells by skipping the stage of dedifferentiation to produce calluses. Some regeneration systems have been established where directly differentiated buds are formed through the induction of explants such as plant cotyledons, leaves, caulicles, hypocotyls, and shoot apical meristems. Such systems experience little somaclonal variation and can well maintain the genetic stability of receptor plants. Transformed exogenous genes can also realize stable inheritance, especially for the directly differentiated bud system formed by shoot apical meristem cells. With a short cycle of obtaining regenerated plants and simple operation, such systems are especially suitable for horticultural plants such as fruit trees and flowers for asexual propagation.

4. Germ Cell Receptor System

The germ cell receptor system realizes genetic transformation using plant germ cells such as pollen cells and egg cells as receptor cells. Currently, germ cells are used for genetic transformation in two main ways: First, pollen cells and egg cells are subject to tissue culture to induce callus cells to further differentiate and develop into haploid plants, thus establishing a haploid genetic transformation system. Second, genetic transformation is realized directly through the fertilization of pollen cells and egg cells, such as by pollen-tube pathway, pollen grain immersion, and ovary microneedle injection. This receptor system allows the exogenous target gene to fully express, which is good for trait selection, and homozygous diploid new variety can be formed after doubling. Therefore, the use of germ cells as transgenic receptors, combined with the haploid breeding technique, can simplify and shorten the complex breeding and purification process.

## III. Procedures for Building Receptor System for Plant Genetic Transformation

The establishment of the receptor system for plant genetic transformation mainly depends on plant tissue culture technique but with higher requirements than in general plant tissue culture. The procedures mainly include the establishment of an efficient regeneration system and antibiotic sensitivity tests.

## Establishment of an Efficient Regeneration System

1. Select the Appropriate Explants

(1) Select juvenile explants where possible. The tissue and organs of a plant are close to the apical zone, shoot tip zone, and growing point zone.

(2) Select explants with strong proliferation ability. In addition to meristems, young inflorescences before meiosis and nucellar tissue after meiosis are explants with generally strong proliferation ability.

(3) Select explants during the germination period. Avoid dormant buds and select those that are about to germinate.

(4) Select explants with strong regeneration capacity. This is because explants of the same plant with different genotypes have different regeneration capacities.

2. Determine the Optimal Culture Medium

All the nutrients required for plant-isolated cells in tissue culture are provided by culture media. A culture medium contains a wide variety of nutrients, mainly including inorganic nutrients, carbon sources, organic additives, and plant growth regulators. So far, many formulas of the minimal medium have been developed. Common in some aspects and unique in other aspects, these formulas should be selected according to specific conditions. The principles for selection include the following aspects.

(1) Explants of the same species, i.e. explants of different subspecies and varieties of the same species, may require the same or similar minimal medium.

(2) Different tissues and organs of the same plant may require the same type of minimal medium.

(3) The nutrients required for plant tissue culture are similar to those for field cultivation.

(4) The concentration of inorganic salts is an important factor in the preparation of minimal media and should be used as an important parameter in the selection of minimal media. The organic components mainly vary in type, while the

content of each component does not vary much.

(5) Sucrose concentration must not be neglected, and the most suitable sucrose concentration should be determined after the type of minimal media is determined.

During plant cell dedifferentiation and redifferentiation, in addition to the nutrients required for growth, the regulation of plant hormones is also crucial. Auxins are important hormones for initiating cell division, which is essential in the process of cell dedifferentiation, callus formation, and rooting. Cytokinins are mainly used to promote cell division and maintain continuous mitosis in already dedifferentiated cells. Auxins and cytokinins have synergistic effects on the growth and differentiation of stem cells. Used together with different concentrations and in different ratios, they play an important regulatory role in cell differentiation. Numerous experiments have confirmed that the concentration and ratio of auxins and cytokinins are the most important means of controlling the growth pattern and organogenesis of plant cells *in vitro*. When plants are cultured *in vitro*, the nutritional requirements and the type and ratio of exogenous hormones vary among plant species and explants, and the culture conditions also determine how plant organogenesis occurs. Therefore, the organogenesis of tissue cultures is regulated by selecting suitable explants and adjusting the medium composition and culture conditions to establish a plant regeneration system.

## IV. Plant Genetic Transformation

The genetic transformation of plants is generally completed through an *in vitro* culture regeneration system. For various plants, this process is basically the same, including the selection of target genes, genetic transformation, and selection of transformants.

The process of plant genetic transformation is as follows: Acquisition of target gene—construction of transformation vector (target gene linked to vector) —introducing target gene into recipient—detection of genetically transformed plants —screening and purification of transgenic plants.

The exogenous genes transferred into the plants include genes for insect resistance, disease resistance, and herbicide resistance, among others for improving crop quality. Based on the type of plant genetic transformation techniques used, the recipients can be leaves, shoot segments, shoot tips, calluses, adventitious buds, embryoids, suspension-cultured cells, protoplasts, floral organs, ovaries, and embryos.

At present, the most commonly used technique for plant genetic transformation is *Agrobacterium*-mediated transformation. The commonly used vectors are generally binary expression vectors, including pBI121, pCAMBIA vectors, and Gateway vectors. All these vectors have reporter genes in them that are easy to detect. Their expression can reflect the transformation efficiency and help simplify the late-stage detection of transgenes.

Vectors for plant genetic transformation generally carry various selective marker genes, mainly antibiotic resistance genes. With the use of such marker genes, non-transformed cells, tissues, organs, and plants that do not contain the marker genes or their products die, and transformants survive because of their corresponding resistance.

In general, the procedures for plant *Agrobacterium*-mediated transformation are as follows.

Step 1: Apply the preserved solution of *Agrobacterium tumefaciens* with binary expression vectors (containing the target genes) in a solid medium plate to activate the *Agrobacterium tumefaciens*. Select monoclonal antibodies for propagation culture, and adjust the concentration of the solution of *Agrobacterium tumefaciens* to a proper level for later use.

Step 2: Prepare the plant material to be transformed, such as sterile plant materials or tissue culture plantlets. Pre-culture the explants (calluses, leaf pieces, shoot segments, and petioles).

Step 3: Transfer the pre-cultured explants into the solution of *Agrobacterium tumefaciens* and shake gently for 20–30 min. Then, take out the plant materials and transfer them into the plant media for co-culture (dark treatment).

Step 4: After co-culture, transfer the plant materials to induction media to induce the differentiation of buds. The induction medium contains antibiotics corresponding to the selection pressure, as well as antibiotics (cephalosporin or timentin) that inhibit the growth of *Agrobacterium*.

Step 5: Continue the culture of regenerative buds in the medium containing the same antibiotics as the induction medium to allow the buds to elongate.

Step 6: Carry out the rooting culture in the medium containing the same antibiotics as the induction medium.

Step 7: Transplant regenerated bacterial-resistance plants and detect positive plants through reporter gene assay, PCR testing, and Southern blotting, among others.

Step 8: Observe the traits of positive plants in the fields, allow them to self-cross, select their progeny, and pick homozygous transformed plants.

## Task 2 Tobacco Genetic Transformation

### I. Preparation of Experimental Materials

1. Preparation of Tobacco Tissue Culture Plantlets

Obtain tobacco tissue culture plantlets using the seeds of the *N. tabacum* cv. Xanthi. Select full-grown seeds. Rinse them with 70% ethanol for 30 s on a clean bench and then with sterile water twice. Disinfect them for 8 min with 10% sodium hypochlorite solution, rinse them 5 times with sterile water, and inoculate them in 1/2 MS medium. After 30 days, select young and flat leaves for genetic transformation.

2. Vectors and Strains

pBI121 vector and *Agrobacterium tumefaciens* strain EHA105 are used. Streak *Agrobacterium tumefaciens* strain EHA105 containing pBI121 vector on LB solid medium (containing 50 mg/L Kanamycin and 50 mg/L rifampicin) and culture them at 28°C for 2 days. Select 2–3 single colonies with good growth on the plate, inoculate them into 50 mL liquid LB medium, and culture them at 28°C by shaking them at a speed of 220 rpm until OD600 value reaches 0.3–0.5 (about 12–16 h). Centrifuge at 5,000 rpm for 8 min at room temperature, remove the supernatant, and invert the centrifuge tube on sterile filter paper. Remove the supernatant as much as possible and then suspend with 50 mL MS liquid medium for later use.

## II. Genetic Transformation

1. Pre-culture

Take tobacco leaves that have been subcultured for 15 days, cut them into pieces approximately 0.5 cm in edge length, place them on MS medium, and culture them for 2–3 days.

2. Infection

Pour the suspended bacterial solution into the pre-cultured leaves, and shake them gently to allow infection for 10–15 min. Pour out the bacterial solution and place the leaves on sterile filter paper to dry them.

3. Co-culture

Inoculate the infected leaves in the tobacco co-culture medium (MS+3 mg/L 6-BA+0.2 mg/L IBA, pH 5.8) and place them in an incubator in darkness for culture at 24°C for 3 days.

4. Screening

Take tobacco leaves that have been co-cultured for 3 days and inoculate them on a tobacco screening medium (MS+3 mg/L 6-BA, +0.2 mg/L IBA+ 100 mg/L kanamycin + 400 mg/L carbenicillin, pH 5.8) to allow the leaf wounds to be in full contact with the medium, and culture them in a lighted culture room at 25±2°C. Bacterial-resistance buds can be differentiated in about 20 days, and the screening medium should be replaced every 20 days.

5. Rooting Culture

When there are sufficient bacterial-resistance buds, transfer them into a rooting medium (MS + 3 mg/L 6-BA, + 0.2 mg/L IBA+ 100 mg/L kanamycin + 400 mg/L carbenicillin, pH 5.8) and detect them after rooting.

6. GUS Histochemical Staining

Cut the apical young leaves of rooted plantlets into 0.5-cm-long pieces, and put them into centrifuge tubes with GUS staining solution in an incubator to stain them at 37°C. After 16–24 h, take out the leaves, wash them with 20% ethanol for 20 min and then with 50% ethanol for 30 min, and finally, decolor them with 70% ethanol until the leaves turn blue (not green at all). This indicates the preliminary success of the genetic transformation.

## III. Related Knowledge

1. Influencing Factors of *Agrobacterium*-mediated Genetic Transformation

The whole process of *Agrobacterium*-mediated genetic transformation through the plant *in vitro* culture is long and involves many steps. In addition to the regeneration frequency of *in vitro* culture, several factors greatly impact on the transformation rate.

(1) Activity of *Agrobacterium tumefaciens*

*Agrobacterium tumefaciens* strains vary in virulence (infection), which can affect the efficiency of plant transformation. To ensure the efficiency of infection, *Agrobacterium tumefaciens* should be cultured until they show strong infection ability, i.e. in the logarithmic growth period. Generally, a bacterial solution with an OD600 value of 0.3–0.6 is used to infect the plant material.

(2) Pre-culture time

The pre-culture of explants is to promote cell division, prepare the receptor cells for integration with specific genes, and facilitate smooth contact between explants and media. Pre-culture time is an important factor that affects the transformation efficiency of explants. Only cells in the S phase of cell division (DNA synthesis phase) may be transformed by exogenous genes. Cell division being too short will affect transformation as the cells are not in the best division phase; cell division being too long will not only affect the transformation efficiency but also increase the probability of false positives.

(3) Infection time and co-culture time

The *Agrobacterium* infection and co-culture of explants are very important as this process involves the attachment of *Agrobacterium tumefaciens*, and the transfer and integration of T-DNA. If the infection time is too short, little *Agrobacterium tumefaciens* will attach, and the transformation efficiency will be low. If the infection time is too long, too much *Agrobacterium tumefaciens* will attach, and it will be difficult to inhibit *Agrobacterium tumefaciens* in the subsequent culture, resulting in contamination and death of the plant cells. Generally, after

*Agrobacterium* attaches to the explant, the strain can induce crown gall tumor only after 8–16 h of survival at the wounding site, so the co-culture time must be longer than 8–16 h. However, the co-culture time should not be too long; otherwise excessive growth of *Agrobacterium* will occur. Infection time and co-culture time are 2 interacting factors, and high transformation efficiency can be achieved only by selecting the optimal combination of both.

(4) Addition of phenolic substances

The transformation of the gene begins with the attachment of *Agrobacterium* to the plant wound, and the wound response is necessary for transformation. Most dicotyledons induce plant cells to secret certain phenolic substances after the wounding of tissues. These phenolic substances are known as wounding signaling molecules. *Agrobacterium* is chemotactic to these phenolic compounds, and these wounding signaling molecules induce the activation of *Agrobacterium*. Acetosyringone is a natural inducer of *Agrobacterium*, and its addition can improve the transformation efficiency in the case of *Agrobacterium* infection.

(5) Appropriate selection pressure

To facilitate the screening of transformed cells and inhibit the growth of non-transformed cells, selective marker genes are generally added to the transformation vector, and corresponding antibiotics are added to the selective medium for selection. In genetic transformation, appropriate selection pressure can enable a preliminary efficient selection, avoiding excessive workload due to excessively low selection pressure or early killing of positive transformed cells due to excessively high selection pressure. The selection of antibiotic concentration, i.e. selection pressure, varies with different plant materials, so antibiotic sensitivity experiments must be carried out before formulating selection plans. An appropriate selection pressure that can just inhibit non-transformed explant differentiation or plant rooting should be selected. In addition, rooting selection of transgenic plants is a key step in reducing false positive results. An excessively low selection pressure can lead to false positive results, and an excessively high concentration can affect the nutrient transfer in the early stage of cell growth, which then affects the normal growth of transformed cells.

2. Assay of Transgenic Plants

After the resistant regenerated plants are obtained, an assay should be carried out to confirm that the exogenous gene is transferred and integrated into the genome of the plant cell. At the initial stage of the assay, polymerase chain reaction is often used to quickly select the transformed regenerated plants from the batch, but there are generally a lot of false positive results for the plants after the PCR assay. Therefore, a subsequent nucleic acid hybridization assay should be performed to verify.

3. Field Trait Observation of Transgenic Plants

Positive plants confirmed to be transformed through assay should be planted in a controllable environment, such as in a greenhouse with an insect-proof screen, and field trait observation should be carried out during the whole growth and development process to analyze whether the phenotype corresponding to the exogenous gene appears. Meanwhile, transformed plants should be conducted with self-cross or hybridization to select homozygous transgenic plants at exogenous gene loci from the offspring. These plants will be used for further analysis.

## Task 3 Strawberry Genetic Transformation

1. Materials

(1) Plant materials

Tissue culture plantlets of the octoploid strawberry Benihoppe are subcultured on a strawberry subculture medium for the experiment.

The common subculture medium for the octoploid is MS+0.5 mg/L 6-BA+ 0.1 mg/L NAA or IBA, with a pH value of 5.8. In addition, the medium should be moderately hard and autoclaved with the amount of agar powder determined according to its hardness, not exceeding 8 g at the maximum.

The growth status of the subcultured plantlets directly determines the success or efficiency of transformation. In general, in terms of leaf age, the leaves of octoploid tissue culture plantlets must not be more than 40 days old and should be green and tender with a flat surface.

(2) Preparation of main solutions and culture media

The media are prepared as follows:

① Strawberry subculture medium.

4.4 g/L MS, 30 g/L sucrose, 5.5 g/L agar, 0.5 mg/L 6-BA, 0.1 mg/L NAA or IBA, pH 5.8, autoclaved for later use.

② Strawberry co-culture medium.

4.4 g/L MS, 30 g/L sucrose, 5.5 g/L agar, 1–2 mg/L TDZ, 0.2 mg/L 2,4-D or IBA, pH 5.8, autoclaved, cooled to about 60°C and filled on the plate for later use.

4.4 g/L MS, 30 g/L sucrose, 5.5 g/L agar, 2–6 mg/L BA, 0.2 mg/L 2,4-D or IBA, pH 5.8, autoclaved, cooled to about 60°C and filled on the plate for later use.

③ Strawberry antimicrobial medium.

4.4 g/L MS, 30 g/L sucrose, 5.5 g/L agar, 1–2 mg/L TDZ, 0.2 mg/L 2,4-D or IBA, pH 5.8, autoclaved, cooled to 60°C and added with 300–400 mg/L carbenicillin; Then dispensed into culture flasks for later use.

4.4 g/L MS, 30 g/L sucrose, 5.5 g/L agar, 2–6 mg/L BA, 0.2 mg/L 2,4-D or IBA, pH 5.8, autoclaved, cooled to 60°C and added with 300–400 mg/L carbenicillin; Then filled into culture flasks for later use.

④ Strawberry rooting screening medium (FMS4).

4.4 g/L MS, 30 g/L sucrose, 5.5 g/L agar, 0.1 mg/L IBA, pH 5.8, autoclaved, cooled to 60°C, added with 20 mg/L kanamycin and 300 mg/L carbenicillin and filled into culture flasks for later use.

(MS used above is powder. If mother liquor is prepared, add it in accordance with the amount of mother liquor.)

2. Operation Methods

(1) Strain Activation

① Streak *Agrobacterium tumefaciens* EHA105 containing plasmid (strain varies from lab to lab, and strain EHA15/LBA4404/GV3101 is commonly used for strawberry transformation) on LB solid medium and culture at 28°C for 2 days.

② Select one well-grown single colony (double-check to make sure) on the plate, inoculate it into 50 mL of LB liquid medium at 28°C, and shake it at 220 rpm (for about 12–16 h) until OD600 value reaches 0.4 (only resistance vectors are added to the LB liquid medium and no other antibiotics, such as rifampicin, are added).

③ Centrifuge at 5,000 rmp for 8 minutes at room temperature, remove the supernatant, and invert the centrifuge tube on the sterile filter paper. Remove the supernatant as much as possible, resuspend the colony with 100 mL of MS liquid medium, and shake it for about 1 hour until the OD600 value reaches 0.4 (this value is very important and must be controlled within 0.4–0.6, preferably close to 0.4). Then, the activated bacterial solution can be used for the next step of plant material infection.

(To improve the transformation efficiency, add some AS or SA into the MS suspension as appropriate.)

(2) Process of Strawberry Genetic Transformation

① Preparation of plant materials.

Take 30-day-old young tender flat leaves of strawberry tissue culture plantlets, remove the leaf tip and margin, and cut the remaining part into pieces about 0.4 $cm^2$ for bacterial infection. Try to make sure that there are wounds around the strawberry leaves to make it easier to induce calluses (the joint between the leaf and the petiole must be cut because the differentiation of calluses and buds is easier and quicker here). Cut the leaves and put them into the MS liquid medium.

② Genetic transformation of strawberry.

Infection: Pour the suspended bacterial solution into the cut leaves, and gently shake to infect them for 40 minutes (shake once every 5 minutes during the process to increase the contact area between the bacterial solution and the leaves). (timing: For cultivated octoploid or diploid seedlings, generally infect with the bacterial solution with OD600 value being 0.4 for not more than 20 minutes. If the infection lasts too long, it will be impossible to inhibit the *Agrobacterium tumefaciens* in the late culture stage. For diploid tissue culture plantlets, infect with the bacterial solution with OD600 value being 0.2 for 40 minutes to 1 hour.) After the infection is completed, pour out the bacterial solution and place the leaves on sterile filter paper to dry up.

Co-culture: Inoculate infected leaves in strawberry co-culture medium and culture them in the incubator in darkness at 25 °C for 3 days. (Co-culture should at least last 2 days. If *Agrobacterium tumefaciens* does not overflow, culture for three days.)

After co-culture, if *Agrobacterium tumefaciens* overflows slightly or seriously, it is necessary to wash the leaves with sterile water added with antibiotics, and the washing time should not be too long.

Antimicrobial culture: Inoculate the strawberry plantlets co-cultured for 3 days in the strawberry antimicrobial medium to allow the wounds to get in full contact with the medium and culture them in the lighted culture room at 25±2°C. After about 20 days of culture, white calluses will grow on the edge of the strawberry leaves. After about 40 days (longer for diploids, usually up to 50 days) of culture, resistant calluses will differentiate into resistant plants. (All leaves and petioles of the strawberry plantlets have good regeneration capacity. Therefore, to improve the transformation efficiency and avoid wasting materials, both leaves and petioles should be used as explants.)

If there are reporter genes in the vectors, a majority of the non-GM plantlets can be removed through reporter assay. The GFP can be directly exposed to fluorescence to detect non-GM plantlets. Those that do not light up are non-GM plantlets, while those that do are GM plantlets. GUS staining: Cut a small piece of tissue from the plantlet to stain it. If it turns blue, it is transgenic.

Some strawberry varieties require culture in darkness for 7–21 days, with 7 days as a cycle. The culture cycle may last 7, 14, or 21 days. No screening is required in this step. Therefore, only antibiotics, cephalosporins, and carbenicillin are required to be added to the culture medium. Attention should be paid to the concentration of such additives, which should be as low as possible as long as *Agrobacterium tumefaciens* can be inhibited. Otherwise, they tend to cause damage to plants.

Rooting Culture: When the resistant plantlets grow to about 2–3 cm in height, transfer them to the rooting screening medium for rooting. To obtain complete resistant plantlets, it is necessary to add screening antibiotics for screening. With such antibiotics added, the plantlets with no roots are non-GM plantlets. However, those rooted are not necessarily GM plantlets. Further detection is needed through reporter gene assay or PCR testing. To test the strawberry fruit, wait until the transplanted strawberry plantlets blossom and bear fruit so as to further determine the transgene stability.

## Summary

Genetic transformation experiments are different from the previous tissue culture experiments although operation methods and techniques for tissue culture have also been used. Since such experiments involve many processes and strict requirements, successful operation in each step directly affects the outcome of the next step. Therefore, students should not make any mistakes during the operation in each step. Otherwise, all previous efforts will be in vain. In this module, students will acquire basic theoretical knowledge, understand the operational procedures, and complete the basic operations of genetic transformation experiments. They will also have an intuitive understanding of the relevant materials, instruments, and procedures involved in such experiments.

## Review Test

1. Briefly describe the precautions during tobacco genetic transformation.

2. Briefly describe the factors that influence the transformation efficiency.

3. Briefly describe how leaf tenderness affects the genetic transformation of strawberry plantlets.